COURTSHIP
in the
ANIMAL
KINGDOM

COURTSHIP
in the
ANIMAL
KINGDOM

(Originally published under the title
The Dance of Life: Courtship in the Animal Kingdom)

Mark Jerome Walters

ANCHOR BOOKS
DOUBLEDAY
NEW YORK   LONDON   TORONTO   SYDNEY   AUCKLAND

AN ANCHOR BOOK

PUBLISHED BY DOUBLEDAY

a division of Bantam Doubleday Dell Publishing Group, Inc.
666 Fifth Avenue, New York, New York 10103

ANCHOR BOOKS, DOUBLEDAY, and the portrayal of an anchor
are trademarks of Doubleday, a division of Bantam Doubleday
Dell Publishing Group, Inc.

This book was originally published in hardcover under the title
The Dance of Life: Courtship in the Animal Kingdom by Arbor House,
William Morrow and Company, Inc., in 1988. The Anchor Books edition
is published by arrangement with William Morrow and Company, Inc.

Library of Congress Cataloging-in-Publication Data
Walters, Mark Jerome.
    Courtship in the animal kingdom
    Mark Jerome Walters.
    1st Anchor books ed.
        p.   cm.
    Bibliography: Includes index
    ISBN 0-385-26338-4
    1. Courtship of animals.   I. Title.
QL761.W284   1989   89-34893
591.56'2--dc20   CIP

For Mariko

Contents

Soft, to your places, animals,
Your legendary duty calls.

—Thomas Kinsella

Preface

"THE OCEANIC FEELING of wonder is the common source of religious mysticism, of pure science and art for art's sake," Arthur Koestler wrote. That feeling is also the source of this book, and it is my hope that readers will be inspired by the same sense of wonder as was the writer.

A discussion of courtship can be a gateway into the broader topic of how the study of animal behavior is conducted, its methods, its horizons, its limitations. But while scientific conclusions are a part of this book, they aren't necessarily its mainstay. Rather, several broad themes form this book's foundation—the mystery of how and why sex began, and the male-female antagonism that resulted, to mention a few. I've approached the topic in this manner because themes often endure, while findings—even those of a scientific nature—constantly change.

Although some sections delve into intriguing, if somewhat abstract, theories about animal courtship, I've tried wherever possible to stay out of the laboratory and remain instead in the field, where the wildlife biologist conducts most of his research. Consequently, this book ranges widely both in space and time, from the beaches of Cape Cod to the streets of Kisumu, Kenya, from the Carboniferous Period, approximately 300 million years ago, to today.

However you cut it, animal behavior covers an incredibly vast area. Even the more focused topic of courtship is too broad to cover in detail in a single book. To bring some sense of progression and harmony to this unwieldy subject, I've organized this book along evolutionary lines. For example, it introduces the subject of courtship by describing one of its most ancient practitioners, the horseshoe crab, and it ends with one of the most sophisticated non-human courtiers, the pygmy chimpanzee.

Finally, there is the question of where humans fit in a general portrait of animal courtship. On one hand, we see ghosts of ourselves—sexual or otherwise—in much of what animals do. Certain animals remain paired for life and defend their mates or their young until death; others "cheat" on each other, kill one another in the course of sexual disputes, and commit "rape." The fact remains, however, that humans occupy an intellectual plane so far above that of other animals that one must be very careful indeed when one compares human behavior to that of, say, the great apes, let alone guinea pigs and iguanas.

While the true relevance of the behavior of other animals will remain a hotly debated topic, this book treats animal sexual behavior as but a narrow window through which we can gain insights into the possible evolution of human behavior, not as a concise explanation for it. Perhaps most dangerous of all is the tendency of some observers to use animal behavior to justify what humans sometimes do—arguing, for example, that because the males of some species lord over the females, this must set a precedent for similar behavior among humans. Such discussions often neglect to poitn out that for every example one might find to support views of convenience, a counter-example could as easily be found. This book, while highlighting some facets of human behavior that seem to parallel those of other animals, doesn't hold animal behavior to be the key to unlocking the mysteries of human activity and enterprise. Far from it, many secrets of human behavior are sure to remain just that—secrets.

Since I am not a scientist, most of my knowledge about the subject comes from people in the scientific community. I would like to express my appreciation to those who spent time with me in their labs, in the field, on the telephone, and in lengthy correspondence. They took time to comment upon relevant chapters, not only in first draft, but in subsequent drafts as well. What's more, they understood the difficulties of writing about science without losing that ocean of wonder at its source.

George Archibald, International Crane Foundation; Robert Barlow, Syracuse University; M.J. Baum, Boston University; Mark Botton, Fordham Univesity; Jack W. Bradbury, University of California, San Diego; Jules M. Crane, Cerritos College, Norwalk, California; Jeremy Dahl, Yerkes Regional Primate Research Center, Emory University, Atlanta; Paul Forestell, Kewalo Basin Marine Mammal Laboratory, Honolulu: Anthony Knap, Bermuda Biological Research Station, St. Georges, Bermuda; Burney Le Boeuf, University of California, Santa Cruz; Mary LeCroy, American Museum of Natural History; Lynn Margulis, Boston University; Vic Mastro, Research Methods Division, United States Department of Agriculture, Otis, Mass.; Claire Mirande, International Crane Foundation; Ron Nadler, Yerkes-Atlanta; John Olguin, Cabrillo Beach Marine Museum, San Pedro, California; George Preti, Monell Chemical Senses Center, Pennsylvania State University; Dorion Sagan, writer; Robert Trivers, University of California, Santa Cruz; Merlin Tuttle, Bat Conservation International, Brackenridge Zoology Field Laboratory, University of Texas, Austin; Kent Vliet, University of Florida; Ed Waltz, Syracuse University; Craig Warren, International Flavors and Fragrance Company, New York; Fred Wasserman, Boston University; Larry Wolf, Syracuse University; Cathy Yarbrough, Yerkes-Atlanta. The contributions of numerous other scientists on whose papers I relied are included in the bibliography.

John Alcock of Arizona State University read the entire manuscript and offered many helpful suggestions. Thomas Clemmons provided several key sources. Olinda Woods researched the derivations of words whose etymologies are included.

Preface

Eden Collinsworth proposed the idea for this book; James Raimes helped shape it from the beginning; and Allan Mayer greatly improved the original manuscript by his skillful editing. My agent Richard Balkin was always there with helpful advice.

Although not directly involved with this project, those who have supported my writing in general are also very much a part of it: Ken Gilmore has offered something for which there is no substitute—encouragement; Jeremy Dole took a chance and gave me my first writing assignment; and Clell Bryant has generously shared his editorial advice.

I am also indebted to the family of Dr. Hideki and Violet Sakurai, whose cheerful company and open invitations to dinner throughout the course of writing saved me from hunger one night, loneliness the next. Finally, I would like to thank my mother, who spent the greater part of parenting driving me to and from the pet shop.

1.

Ancient Beginnings

THE EARLY CARBONIFEROUS PERIOD some 300 million years ago must have been almost paradisiacal in richness and tranquility. There were no snakes or apple trees, and no large animals made noises in the woods at night. Only a few insects had evolved sufficiently to live completely out of water. Lush seed ferns grew along rivers that moved with slow certainty, pole-straight trees crowned with ferny growth grew one hundred feet tall, and opalescent-winged dragonflies rode humid breezes through the greenery. The forerunners of horsetails grew along the marshes, and the water was full of exotic trilobites, creatures resembling horseshoe crabs. Hidden behind a humid atmosphere, the sun was never visible as a distinct orb, but passed over the earth in a shower of yellow light, while silk curtains of rain swept over the land. Carboniferous forests, whose ultimate death, burial, and biological decomposition produced the coal and gas that, hundreds of millions of years later, would fuel the great civilizations of earth, were just beginning to appear. The debut of dinosaurs was still thousands of millennia away.

Despite the distance in time and imagination, what was happening back then has everything to do with who, what, and where we are today. Evolution had already made some basic career decisions about life. Animals had begun to break their bonds with the sea that stretched back to the beginning. Life was in an expansive mood, and, consequently, there was already a high premium on real estate. Many species crowded along the shores like refugees in an overpopulated country, waiting for a quirk of evolution to let them out. But perhaps the most fascinating change was that, instead of simply dividing in half—or cloning—in order to reproduce, as bacteria and similar organisms had done since the beginning, life had developed something known as sexuality. Instead of one creature always cloning into two, two creatures could fuse to become one. There were "males" and "females"—and in some cases, other sexes too. Some of today's paramecium still have many mating types, which are designed by letters of the alphabet or Roman numerals. A can mate with B, for example, but not with another A. With this drift toward sexuality, the tiny machinations of courtship and mating among animals already hinted at the appearance of connubial bliss, sexual enmity, and Adam and Eve.

This, of course, did not happen overnight. For some two billion years, reproduction had required neither partners nor sex. Individuals simply split in two, as the bacteria still do. But there was sex—of a sort. In a strict biological sense, bacteria practiced sex when one of them simply gave genetic material to another. In some cases, it involved two cells nudging up to one another, and momentarily joining, or suspending a tube between them, and one donating "his" gene to "her"—a process known as conjugation. The two cells then parted, forever changed by the experience. But rather than a new individual being born as a consequence of this union, the recipient organism was simply given a genetic facelift. Actual reproduction had to wait for the subsequent cloning of one of the remodeled cells. But that was another story altogether. Sex and reproduction were independent events—a distinction that we humans have a hard time appreciating.

Then, some 500 million years ago, long before the appearance of the first animals, a momentous change began to occur in the history of life. After a long line of virtually imperceptible changes in individuals, visibly different organisms evolved. A new kind of sexuality had emerged: it involved two individuals meeting, not to exchange genetic material, but actually to "mate." Each parent contributed genetic materials that would fuse and become the core of an entirely new individual. Sex and reproduction became linked. The genetic material came in two separate wrappings: half would eventually be borne within mobile cells with whiplike tails—the beginnings of sperm; the other half would become sequestered in passive cells—the beginnings of ova. While the parents remained the same, their offspring were genetically unique. This was the beginning of male and female, fertilization, and other traits we associate with sexual reproduction. In terms of the history of life, this was indeed a late chapter, for nearly five-sixths of this history has been written by microscopic, asexual organisms. Today, in fact, the majority of organisms in four out of the five animal kingdoms still reproduce without sex.

But because this new type of sexual reproduction allowed for the joining of vast amounts of genetic information—the biological data from which the new individual would be created—it had the potential to create highly complex offspring. In this sense, from a sort of swampy dark age in which simple microorganisms sheared themselves in two, a biological renaissance began. The reproductive cargo of two individuals, united, could produce an energetic bundle of life capable of growing into something with many kinds of tissues and structures. In short, egg and sperm could hold enough genetic information to produce a worm, mosquito, rabbit, or human being.

The reason behind the momentous appearance of two-parent sex remains among the most contentious questions in evolutionary biology. After all, for at least two billion years life thrived without this kind of sex. Every hospitable place on the earth teemed with

7

asexual organisms. As illustrated by many of today's flourishing asexuals—bacteria that make soil fertile by decomposition or render air breathable by producing oxygen, not to mention those responsible for bubonic plague, botulism, diphtheria and numerous other reproductive success stories—it is possible to do just fine without sex as we know it. Cloning in quiet obscurity, bacteria have moved from the primal swamp straightaway into the modern home and operating room. Some bacteria can reproduce so rapidly that, given sufficient space and nourishment, one of them dividing every twenty minutes could theoretically create a mass greater than the earth's in less than two days. With such impressive credentials, who needed sex?

The traditional explanation has been that two parents became involved in reproduction because they produced highly diverse offspring capable of adapting to new and difficult environments more successfully than their asexual brethren, who were exact copies of their "parents." It has long been argued that except for the occasional useful random mutation—the lucky break—few clones are born with the genetic innovations necessary to survive in changing circumstances. And even if their immediate environment doesn't change, clones may become geographically isolated because there are few pioneers with the genetic fortitude necessary to stake out new territory. With the advent of new-style sex, however, change no longer had to wait on the random mutation. Because each parent stirred its own genes into the recipe, every offspring was unique. Sex made for a sort of instant variation. A mutationlike leap was accomplished at conception. Every individual was different from the next, and consequently had a unique potential to adapt to a new environment. Sexually reproducing species—so the theory goes—could leapfrog over their slower asexual predecessors. Evolution was speeded up.

In fact, there is little evidence to support this long-standing view. Bacteria had mastered their own way of creating genetic diversity through genetic donation. If anything, in many cases asexual organisms—from bacteria to dandelions—are just as successful colonizers as sexual ones.

Of course, it is easy to claim superiority for an invention if one considers only its advantages. But when biologists began adding up the disadvantages of sex, they began to wonder. For one thing, sex is incredibly expensive for the individual. A clone passes 100 percent of its genes to its offspring—earning a 100 percent return on every reproductive investment. In sexual reproduction, each parent is able to contribute only half its genetic material to make way for the other parent's genetic contribution. Given that an organism's ultimate objective in life is to preserve itself by passing on its own genes, this fifty percent tax on every transaction is a steep price to pay. Each parent's contribution will erode drastically with each generation. For example, a child develops from one-half of each parent's genes; a grandchild from one-fourth. After nine generations, fewer than one in every 415 million genes is a direct offshoot of the original parents. Successive generations dilute the contribution at a geometric rate, rapidly driving the genes of the ancestral parents from the picture. The price of sex is great. We dwindle. Scientists still struggle with the logic behind that evolutionary course.

One of the early unorthodox explanations for why this came about was proposed in the 1950s by Professor L. R. Cleveland of Harvard University. He suggested that animal-style sex began not because it was necessarily superior to cloning, but quite by accident—the result of cellular cannibalism. Early organisms, happening upon hard times, began to devour one another rather than starve. The "undigestable" genetic material of the victim was incorporated by the aggressor. Dangerously overloaded with twice the normal amount of genetic material, most of these cells died. But some cells, through random mutation, managed to split in two and survive with the genetic material thus halved. Two visibly different needs—to eat and to reproduce—became intertwined. According to this scenario, two-parent sex evolved as the necessary side effect of the need to eat. In time, the solution would prove more cumbersome than the original problem. In and of itself, sex is a biological waste of energy and time. Although Cleveland actually put together photographs documenting the sequence of cannibalism and the subsequent fusion of genetic material, other scientists were skeptical that this act of one cell devouring another actually laid the evolutionary foundation for sexual reproduction. As one of Cleveland's present-day proponents, evolutionary biologist Lynn Margulis of Boston University, puts it, "Cleveland, with his Misssissippi accent and idiosyncratic ways, never convinced his colleagues."

One of the more recent explanations for the rise of two-parent sex was developed by William D. Hamilton of Oxford University. The basic premise of his theory is that larger, slower-evolving organisms use sex in order to keep evolutionary pace with the faster-evolving parasites that are constantly looking for opportunities to invade them. This novel idea is as follows.

In the beginning, before the dawn of true multicellular organisms, cells massed together into colonies for the purpose of protection and for procuring nutrients. Hamilton called these colonies "friendly . . . building blocks." In response, the ever-opportunistic "protoparasites," or insidious "foes," sailing across the primal waters in their tiny transparent cell casings, would masquerade as "friends," trying to incorporate themselves into colonies and exploit them for food. The cell colonies were forced to evolve a way of distinguishing between friend and foe—that is, by inventing chemical passwords, or defenses, against the invaders. A friend could latch onto the colony; a foe was rejected.

But the protoparasites, through a random, trial-and-error shuffling of genes, eventually duplicated these genetically controlled passwords and broke through the biological defenses. Each time the parasites penetrated the defenses, the colonies had to shuffle their own genes to create a new one. But because the smaller and more mobile protoparasites reproduced more quickly and therefore evolved faster than complex colonists, the colonies were at a great disadvantage. They needed a way to compensate for their slow pace of evolution, to come up with new combinations at least as fast as the parasites could crack them. Sex, because of the rapid new combinations it produces, became their salvation. Through a genetic partnership of individuals, the colonies could rapidly create new and diverse offspring. Ultimately, the parents themselves would fall to the invaders, but not before producing a new generation of offspring with defenses that microbes would have to spend another generation trying to penetrate—or so the latest theory goes.

But the real truth of exactly why sex has become the all-consuming biological activity of so much of animal life remains a mystery. Most hypotheses, after all, are little more than educated guesses, which in the end may prove to be as fanciful as the abundance of myths that sought to explain the appearance of sex. Years ago in Germany, before science explained the fertilization of corn as the result of pollen falling on the stamen in a budding ear, peasants hearing the wind wafting through the fields of wheat would say, "Here comes the Corn Mother." Wind was the giver of life. Similarly, an Indian legend from the southwest United States tells of a maiden who conceived by standing in summer rain. The Pawnee Indians believed that the moon was female, the sun male, and they joined to give birth to the first human. In the cosmology of the Zuni, the Sky Father lay across the Earth Mother, impregnating her. Humankind then emerged from the earth's womb, which some tribes believed to be the Grand Canyon.

An Afro-American legend explains how a society of women split into two sexes. A spider lived in a hollow tree trunk between the village and the river, which was the villagers' only source of

11

water. As the women walked to the river and stepped across the log, the spider impregnated them, sending a shiver through their buttocks. The women later gave birth to children who had something "unnatural" between their legs. A society of males and females was thus born.

Sexual reproduction is actually the story of two unions. In many cases, a male approaches the female, or vice versa, and after a web of subtle communication known as courtship, they mate. In the second union—conception—sperm approaches egg and, following complex chemical reactions, they fuse. Courtship is the bringing together of individuals. Conception is the bringing together of gametes.

For whatever reasons sexual reproduction arose, it seemed to create as many problems as it solved. Now individuals had to coordinate their lives. Male and female first had to find each other, then release gametes with a precise timing to ensure fertilization. Such split-second precision was especially important to the early organisms, for many of them lived in the sea, where the fickle environment would quickly carry away eggs and sperm. Seconds could mean the difference between fertilization, or gametes floating away in the current. Released with near-perfect timing, the roe and the milt, suspended together in the water, would join to form *zygotes*—Greek for "yoked together"—which, with luck, would develop into adults.

External fertilization was convenient for the parents but rough on the offspring, who were left to drift or sink according to the vagaries of the sea. It was every zygote for itself. One way parents compensated for this vulnerability of tiny offspring was to produce prodigious numbers of them. A biologist once determined that a forty-four-pound ling, a member of the cod family, carried some 28,361,000 eggs. Of these, only one in a million has to survive in order to keep the species in good standing. At first sight, this seems a profligate strategy. But the alternative to abundance—to lay a few eggs and then remain around to protect them—is equally costly. Many of the fish and other animals who practice external fertilization have opted for abundance without parental responsibility.

The elaborate courtship of some fish involves a complex choreography that not only lands partners in the right place, but triggers the release of egg and sperm in the required split-second synchrony. The complexity of their "dances" is often astonishing.

At two hours before sunset, a small Atlantic coral-reef fish known as a black hamlet abandons its solitary daytime routine of foraging and defending its territory and goes out searching for a mate—often seeking a familiar partner. The one fish is greeted by the other with an elaborate display or a mock chase. The male, his body quivering, aligns himself in front of the female and quickly dashes forward, braking his approach by flaring out his pectoral fins. As he pitches forward, he raises his tail in full view of the female. He may continue to flare his fins and arch his back like a taut bow. The female, meanwhile, begins to reciprocate.

As they approach to mate, both rapidly snap their heads side to side. The female slowly arches her body, quivering mildly, as if a chill is running up her spine. The male swims up from behind and nuzzles his head between her anal and caudal fins. Then they both slowly curl their bodies together, until the nape of each is under the tail of the other. The female forms an "S" shape, while the male, his jaws clinched, curls tightly around her. The female releases eggs while the male emits milt into the tiny hollow formed between their entwined bodies. The fish then disengage and dart back to the reef, leaving the fertilized eggs to sink and lodge into crevices in the coral.

The most remarkable trait of the black hamlet is that each individual is actually both male and female, or hermaphroditic. After a short rest, when the fish come out again, they reverse roles.

13

The former "male" provides eggs, while the former "female" contributes sperm. They alternate roles throughout the evening, and after spawning perhaps a half-dozen times, return to their respective territories by nightfall.

When animals left the sea, somewhere in the late Carboniferous period, they were deprived of the wet environment required by sperm and egg. So instead of depositing sperm and egg externally into water, many land animals adopted a means of internal fertilization, depositing gametes in a sequestered internal environment. Reptiles encase the fertilized egg in the watery interior of a shell. The animal within, conceived and developed in fluid, then grows out of its watery environment and makes its debut on land. Mammals evolved a uterus—that small internal sea, with tides and a capacity to nurture life. This protective enclave, rich in a network of vessels to nurture the embryo, made the calcium casing developed by reptiles unnecessary.

The move from sea to land had great implications for courtship. Whereas in the sea, partners merely had to be brought into proximity with each other, on land actual mating was usually required. This meant major changes not just in courtship patterns, but also in the physical design of the animals themselves. A simple vent for expulsion of egg or sperm would no longer do.

Amphibians, which inhabit the moist intermediate realm between land and water—shady woodlands rich in humus, brooks with their fern-fringed banks—often use means that represent a middle ground between external and internal fertilization. The salamander employs a tiny packet of sperm called a "spermatophore." The male deposits this protective sack on the ground, and in a precise courtship dance leads the female directly over it so it attaches to her vent, or cloaca. Different species have developed variations on this theme. The North American dusky salamander deposits the sac and, as he moves away, leaves a thin mucous thread. The female, guiding her chin along it, is led directly over the sperm-filled sack. The brook salamander entwines the female with his tail and, in a process that may last two days, attempts to press his vent against hers and deposit the spermatophore directly into the female. During the mating season, the male two-lined salamander of North America grows special fangs. While holding the female from above, he rubs a special "hedonic" gland over her back. He then lacerates her skin with his teeth so the aphrodisiac seeps into her bloodstream. Afterwards, he is able to deposit the spermatophore into the compliant female's vent.

Of the 800 known species of frogs, all of which are amphibious, only the tailed frog, resident of the swift streams of the northwest United States, employs internal fertilization. In such an environment, external fertilization would be highly unreliable. The male introduces semen directly into the female with a primitive forerunner of the penis—a tail with a groove running along it, which unfolds from his genital opening. The male of this primitive species also has the distinction of being voiceless. Instead of calling and awaiting the arrival of females, as do most frogs, he must seek them.

The first group of animals for which an external organ became standard were the reptiles, who descended from the amphibians sometime during the Triassic period. With one exception—the lizardlike tuatara of New Zealand—all reptiles possess an extruded penis. The tuatara, the most ancient reptile in existence, mates in the fashion of salamanders—rubbing against its mate to deliver sperm. For other reptiles, the extruded penis means that mating requires not only the touching of partners, as in the tuatara or salamander, but actual joining of male and female, as in the case of the American alligator.

The courtship of the American alligator has inspired its share of myths. It has been said that the male rolls the female onto her back before mating, and then rolls her over again afterward. It has also been claimed that the animals mate year round. Actually, at least around St. Augustine, Florida, they mate between the second week of April and the first week of June, with peak activity usually occurring about the third week of May.

"The courtship of alligators is leisurely, usually involving a great deal of touching, bumping and pushing on neck and face," says University of Florida biologist Kent Vliet, who has studied *Alligator mississipiensis* in the St. Augustine area for several years. "Once two animals have come together, their courtship may last several days. Often one nudges its snout under the head or neck of the other. There is also a great deal of vocalization, leading at times to an alligator chorus—a communal bellowing that usually occurs early in the morning. Sometimes males engage in very loud and impressive bellowing duels. This not only draws opposite sexes together but may also be territorial warnings to other males. The females have a higher pitched roar, so the males have no difficulty distinguishing between the sexes.

"During these bellowing sessions the male often performs a 'water dance' by sending out very powerful subsonic signals from somewhere in the center of his body. This causes water droplets to rise around him ten or twelve inches into the air like a fountain. His signals are of such low frequency but high volume that they could probably travel tens or even hundreds of miles across a still body of water. When a female detects it, she sometimes stops bellowing, rushes to the male, and puts her snout under his chin, then closes her eyes and seems just to experience the vibrations. The water dance appears to be very stimulating to her. She is selective in the male she chooses. If an unwanted suitor approaches her, she begins to mew and cough, drawing the attention of other males.

"When mating, the male and female lie on the surface with their bodies parallel but facing opposite directions. One turns and rides up on the head of the other and forces it under water— probably a test of strength. Copulation itself is a very difficult se-

16

quence because their vents are on the underside of the base of their tails. The male clasps the female with his front legs, rolls to his side, and slides partially beneath her. They mate for about thirty seconds—quite rapidly compared to reptiles in general."

The evolution of this sort of internal fertilization entailed not only major changes in physical design and courtship behavior, but enormous changes in the relationship between the sexes. Among externally fertilizing sea creatures, gametes are liberated from both parents, who are then free to leave. But when this watery environment moved inside the female, while males continued to liberate themselves of sperm, females became captives of their young. The male gained an exploitive edge in sexual matters. He could come and go; the female could not. In many cases, the female consequently had to develop a way of carefully selecting a male who would offer good parental care. Courtship, far from simply bringing male and female together to mate, became a means of testing his suitability. In short, the change from external fertilization in the sea, to copulation on land, had rewritten totally the story not only of courtship, but of mating and birth as well.

It is one of the traits of evolution that for all that life changes, nothing is totally forgotten. Evidence of our primitive legacy is everywhere. Symbolically and literally, memories of the sea in which we evolved are very much with us. Our own blood, for example, combines the elements of sodium, potassium, and calcium in the same proportions as they are found in seawater. Man is two-thirds ocean, one-third land, so to speak. We are reminded of our own ascent from asexual organisms every time a fertile ovum divides in the human uterus and produces clones—better known as identical twins. Such primal legacies, of course, reach beyond reproductive matters. Six percent of human newborns retain vestiges of a tail—a little known occurrence since evidence is easily removed by surgery. In 1926 the case of a twelve-year-old boy with a nine-inch doglike tail was documented.

One of the most indelible legacies of our past is that, despite the passing of three billion years, animal sex remains a pact between tiny gametes—eggs and sperm. Whether a tiny mouse or a huge blue whale is involved, fertilization requires the union of single cells. Through a fragile and complex chain of events reaching back nearly to the dawn of sex itself, courtship makes this possible. In a world teeming with millions of species and billions of animals, courtship is nature's way of insuring that the right sperm finds the right egg at the right time.

2.

Three Lunar
Dances

IT IS EARLY MORNING on Delaware Bay. A low, steamy vapor lies offshore. The tide, drawn high by the position of a hidden new moon, peaks. You look into the dark sky, a few stars still glimmering faintly along the cosmic ribbon of the Milky Way, and for a moment you almost feel the planetary motion of earth—not a spinning or whirring, but a sensation of slowly reeling through space. For a moment only the tide stirs and the earth rolls. Then a tern suddenly breaks cover in the grass-covered dunes at the beachhead to test its quick, shallow call against the coming dawn. And the moment is gone.

As daylight breaks, thousands of birds flock noisily to the shore—plovers, sandpipers, ruddy turnstones, sanderlings, red knots, willets, dowitchers, and black-headed laughing gulls. Hundreds of thousands of them cover the beaches in mottled brown carpets that slide this way one moment, that way the next. Frantic sanderlings sprint before the hem of every wave washing ashore. Seagulls, standing a neck above the rest, wait for a cache of food to be left unguarded.

Every June legions of birds, many from as far away as Chile

and Argentina, visit New Jersey's Cape May in search of eggs laid by horseshoe crabs. Massing near shore at high tide, the female horseshoe crabs wallow shallow depressions in the sand, lay up to 80,000 eggs each, then catch the outgoing tide. Coinciding as it does with the 6,000-mile journey of birds to their Arctic breeding grounds, the spawning of the horseshoe crabs provides a vital mid-trip refueling.

By 9 A.M. the tide has receded. At about the same time, Marine biologist Mark Botton, a clipboard in hand, steps from a weathered building, an extension laboratory of Rutgers University. With the temperature already in the eighties, Botton wears tennis shoes, shorts, T-shirt, and a sun visor. Although birds still actively forage, most of the horseshoe crabs have returned to deeper water. Others have been stranded on shore. Botton gazes along the beach where thousands of animals lay. While many will survive until the next tide by hunkering down in the sand until the edge of the hard shell rests just below the surface, forming an effective heat shield, thousands of others will perish.

Botton walks across the dunes and down to the tideline. Here and there a horseshoe crab has been flipped onto its back by gulls, who hammer at the soft underbelly with their stout yellow bills. Botton picks up a large carcass and measures it with his tape. "Thirty centimeters," he says. About twelve inches across, it's huge—the size of an elephant's foot. He tallies the victim on a chart, then tosses the carcass onto a sled which he pulls along with a yellow nylon rope. When the sled is full, he dumps the subjects in a nearby graveyard of other shells so as not to confuse them with victims of the next round of mating.

Despite the apparent defeat of horseshoe crabs at the beaks of millions of shore birds every spring, the crabs will prevail. Hardy, versatile, of simple mind but durable design, they live in water both hot and cold, clean and dirty. Since its appearance some 200 million years ago, while millions of other species have become extinct, the horseshoe crab has hardly changed. With as many as fifty nests in every square yard, the crabs can afford to lose most of their eggs to birds. Abundance is insurance against heavy loss.

The horseshoe crab may be the most extensively studied tidal animal in the world. As early as the 1930s, scientists were intrigued by its large optic nerve. Later, its blood was found to possess a unique tendency to clot when brought into contact with minute amounts of biotoxins. As a result, pharmaceutical companies have used its blood for decades to test drugs for biological contamination in drugs. Nonetheless, in many ways, the animal is still a stranger to science. Little is known about how it times its arrival on shore with high tide—the opportune time for breeding. Almost nothing is known of its courtship or behavior on the bottom of the bay, where it spends most of its life.

Limulus polyphemous is not actually a crab but an arachnid, one of a group of primitive animals that includes scorpions and spiders. Ancient and trilobitelike in appearance, it has a smooth, two-sectioned shell. A hard spinelike tail is attached at the rear. Indians who once lived on what is now Cape May used the sharp tail spines for spear tips. The horseshoe crabs use them for righting themselves if flipped by the breaking waves. *Limulus* breathes through thin, overlapping membranes, called "book gills," located beneath the rear of the shell. They propel themselves by simultaneously stroking all eight pairs of legs, which sends them scurrying across the ocean bed in a kind of carefree lunar motion. Perhaps their most interesting anatomical feature is their two pairs of obvious eyes—two large lateral ones, shaped like horizontal tear drops, and a second, much smaller pair at the front of the shell. Barely larger than pinheads, these tiny eyes detect only ultraviolet light and may serve to adjust the sensitivity of the lateral eyes to changing light conditions, like the automatic light meter on a camera. In addition, there are six additional clusters of light-sensitive cells that qualify as eyes—three clusters on top of the animal, two

Mark Jerome Walters

underneath, and one on the tail, which keeps track of day lengths and probably provides an important clue for telling the animals when to come ashore to mate. All told, the horseshoe crab has ten eyes—impressive equipment for an animal often described as primitive.

Time ashore constitutes a relatively brief moment in the life of the horseshoe crab. After a winter on the bottom of Delaware Bay or in the nearby Atlantic, the animals begin moving toward the beaches. By this time, most females already have a male attached to the rear of their carapaces with rounded, boxing-glovelike claws, leading scientists to speculate that pairing occurs in winter. Rather than pairing off every year, it is possible that a male stays locked on for years. "In addition to this prolonged pairing, it's also curious that the female seems totally unable to jettison a male once he's attached," says Botton. "But if the male has total say in the selection process—an unusual situation among animals—females have total say in where they go."

In fact, with males outnumbering females by as much as five to one along the bay, only about twenty percent of males find a mate. The unpaired bachelors form "stag lines" for miles along the shore—a sort of gauntlet of shells—to intercept the mated pairs as they move toward shore. As the bachelors follow the pairs to the beach, dozens of these so-called "satellites" pile on top. The female, buried beneath the onslaught of males, releases eggs, while her attendant male as well as the satellites dowse them with milt. The primary male has a great reproductive advantage, although milt from some satellites might sometimes find the mark. The pair then covers the eggs with sand, mixing in pebbles and bits of debris, in an attempt to conceal at least some of the eggs from the birds, which frantically search the beaches for caches of eggs during ebb tide. Females often lay up to a half-dozen clusters in a straight line along the beach, burying them at different depths, probably in an attempt to foil the birds.

22

It is unlikely that anyone knows more about the mating behavior of *Limulus* than Robert Barlow, a professor of neuroscience at Syracuse University. For the past five summers at the Marine Biological Laboratory, Woods Hole, Massachusetts, he has studied the animals both in controlled lab conditions and on the beaches of Cape Cod. One of his objectives is to determine what triggers their mass migrations to shore—a question he has contemplated while sitting on the bottom of the bay in SCUBA gear. "Great numbers of them waited idly off shore, half buried, looking like a mass of lifeless shells littering the bottom," he says. "Suddenly, on cue, many of them oriented toward shore and began marching past me. I spun around to see what could have prompted them. Nothing obvious. As a human being, I wasn't privy to that mysterious signal.

"What is most uncanny is that they always manage to ride the highest tide of the day to shore. Without the Geodetic Survey tide tables, it is a very complicated business for a human being to determine whether the first or second tide of the day will be higher. But horseshoe crabs get it right every time. What clue do they use?"

Like many tidal animals, the horseshoe crab can predict the tide by somehow cuing in on the phases of the moon, upon which the tide depends. Barlow's discovery that *Limulus* possesses this dual-tide rhythm was an impressive scientific feat, even though the physical mechanism by which the crab distinguishes between the two high tides each day remains a mystery.

But there was a second question that also puzzled Barlow. Not only did *Limulus* manage to make it to shore at just the right time, but the unattached males seemed to have an uncanny ability to locate females, even in darkness. "Scientists who knew a lot about them warned me, 'The crab's eyesight is useless for mating, so if you're looking in that direction you're wasting time,'" Barlow says, standing in his laboratory, surrounded by an array of oscilloscopes, centrifuges, electronic timers, digital timers, amplifiers, and computers. "Yet I knew from tests that *Limulus* had excellent vision." Convinced that they indeed used their eyes, Barlow devised a simple experiment to test the theory.

As a start, he cast some cement horseshoe crabs, using an empty shell as a form, painted them various shades ranging from white to black, then planted them along the shore at high tide, in daylight, when the crabs migrated in to mate. He then watched to see how the males responded. For weeks he collected data. "I was on the verge of completing the experiment when, one morning, I returned to the beach and found that all the *Limulus* had disappeared," he says. "There had been thousands the day before, and now the shore was almost barren." Barlow suspected that fishermen had plundered the beaches during the night and had sold his subjects to laboratories that extracted their blood for its toxin-detecting qualities. He began inquiring about "who took my *Limulus.*" On a tip, he went to a fisherman's house and spotted thousands of them piled under a tarpaulin in the backyard. Barlow paid the man $435 for 1,200 of them, which he took back to the beach. But their will to mate had been so depleted by the ordeal that it was futile to continue. Through subsequent experiments, however, he did confirm that in daylight, males readily spotted black and gray dummies, which were virtually indistinguishable from real females. His theory seemed to be holding up.

Data from the laboratory also gave Barlow reason to suspect that the horseshoe crabs had keen night vision as well, suggesting that a good part of their courtship probably occurred then. In fact, laboratory experiments confirmed that at night their vision actually became up to 100,000 times more sensitive than during the day. Barlow then began conducting the dummy experiments at night. But observing the crabs in near total darkness posed an obvious problem: how would he see them?

Barlow invested a portion of his National Science Foundation grant in an infrared camera. His plan was to place several dummies in the water, set up the camera, and video tape bachelors that passed by. He could then play back the video tape in his lab and chart the course of each male. But he soon discovered that plankton in the water blocked most of the infrared light, and what little did reach the crab was reflected, not absorbed, by their shells. Ghosts appeared on his monitor screen. The image was not much better than what he could see in darkness with his own eyes—almost nothing.

In 1986 Barlow happened upon a technological savior that brought new hope to his struggle to understand the nighttime courtship of *Limulus*—a military instrument known as an image intensifier, which could practically turn night into day by amplifying the few available photons of light. Equipped with this costly new instrument, Barlow set out nightly, week after week, into the nocturnal world of the horseshoe crab. On one such outing, I went along.

The weather that evening made the earth seem very old—dank and cold, with localized rain squalls moving low across the gray horizon. Barlow's new white Jeep, with NATIONAL SCIENCE FOUNDATION–SYRACUSE UNIVERSITY FIELD RESEARCH signs on each door, pulled out of the grassy drive of his summer residence directly across the harbor from the Marine Biological Laboratory. Above the warm hum of the heater, Beethoven played on the radio as the wipers slapped at the mist on the windshield. The electronic equipment inside the vehicle gave it the feeling of a modern tank: a four-inch monitor, cables, a video camera with an image intensifier attached, and a portable VCR between the front seats. An eighteen-foot aluminum ladder was attached to the roof.

When we arrived at Mashnee Dike—the site of most of his experiments—Barlow parked the Jeep near the tide line. We slid the ladder off the roof of the vehicle and bolted it like the boom of a crane to a custom-made hinge on the front fender. We attached the image intensifier to the lens, lashed the camera to the ladder's end, which we raised ten feet off the ground. Barlow then placed a dummy at the water's edge, and maneuvered the Jeep until the camera hung directly over it. We huddled in the cab against the drizzle and turned on the video monitor and recorder.

When bachelor males spotted the "female" dummy—even from as far away as ten feet—they made sudden, determined changes in direction. They nuzzled the dummy with the front of their carapaces, courted it, tried to mate with it, and sometimes even released sperm as they butted it. Only after repeated rejections would a male call it quits and head back to deep water. New suitors arrived continuously, scurrying toward the circular dummy. Miles of video tape, meanwhile, were being recorded. Later it would be played back, the course of every bachelor charted, until the acuity of their night vision had been established. While many hours of analysis remain, it is clear that the horseshoe crab can spot a "dummy" at night from a far greater distance than can a human. The experiment seemed to confirm what Barlow had suspected all along: *Limulus* not only used their eyes during courtship, but could locate a mate as easily at night as during the day.

The most remarkable discovery was not that their eyes served them well in mating, but that the sensitivity of the eyes followed a solar rhythm—that is, they grew extremely sensitive every twelve hours or so. In effect, Barlow had discovered a clock-within-a-clock in the horseshoe crab. There was a dual-tide rhythm by which *Limulus* migrated to shore at the opportune moment for mating, and another twenty-four-hour cycle, tied to the earth's rotation, that determined the sensitivity of its eyes. Barlow suspects there may be yet more rhythms by which *Limulus* lives. But what he has confirmed thus far is astounding, especially given that the horseshoe crab has not seen a major external design change in over 200,000 millennia.

Though it didn't seem possible, the night grew darker. By 10 P.M. only the splintered porch lights of homes across the bay shone dimly through the rain. Barlow shrugged his shoulders. The weather was bad and, despite the new moon, disappointingly few animals had appeared. Then the image intensifier failed, a back-up battery died, and a cable was pinched in the hinge between the bumper and the ladder. Finally, the ladder fell on a research assistant's foot, bruising it. "Nobody said field research is easy," Barlow said with a sigh. "If it's not one thing it's another." His Syracuse University Swimming jacket (an advisor to the swimming team, he sometimes swims ten miles across Buzzard's Bay to New Bedford) doused with cold rain, he helped disassemble the rig, climbed into the Jeep, and we headed home.

Courtship is one of many animal behaviors orchestrated by biological rhythms. Numerous bodily processes and sensitivities follow cycles. A loud noise that can spark a fatal convulsion in mice at one time of day hardly fazes them at another. Doses of X-rays that will wipe out a population of mice at one time, kill only a few of them hours later. Even the susceptibility of insects to insecticides partly depends on time of day.

Like a house filled with different clocks, organisms abound with devices that tell them when to eat, sleep, and reproduce. Some clocks are set according to the sun or the moon. From birth to death, they largely chart the course of the lives of animals—humans included. The peak time for the onset of labor, for example, is between 1 A.M. and 7 A.M. The human's favorite time to die is about 6 A.M.

When a biological clock is set to the sun and runs on a twenty-four-hour solar day—the daily transformation of the eye of *Limulus* is one such clock—it is described as having a circadian rhythm (from Latin *circa*, meaning "about," and *diem*, "day"). Circadian rhythm influences a wide range of biological responses

27

and behaviors. The adrenal gland, brain, heart, kidney, liver, pancreas, pineal gland, skeletal muscles, spleen, and chemical composition of blood and urine all follow twenty-four-hour fluctuations. Even human sensitivity to house dust, which on average peaks after 11 P.M., seems to follow this drummer.

While a circadian rhythm is calibrated to the solar day, other biological rhythms are set to the year. These are known as annual rhythms. A dramatic case of annual rhythm is that of the golden-mantled ground squirrel. Just before the onset of winter, it lowers its body temperature to a degree or two above freezing, drops its heart rate to one beat a minute, and falls into a deep sleep.

Somewhere between the tick of circadian rhythm and the tock of annual rhythm falls a beat known as the lunar month. This is the 29.5-day period between new or full moons, or the time required for a full rotation of the moon around the earth. In addition to the lunar month, there is also the 24.8-hour lunar day, the period between moonrises, or the time it takes for the earth to make one rotation on its axis in relation to its nearest neighbor.

Despite the slight time difference betewen the lunar and solar day, what happens when a human being apparently falls out of step with the latter can be dramatic. Take the case of a blind man referred to as J.X. by the scientists who studied him. J.X. *seemed* to have fallen in step with the 24.8-hour lunar day instead of the normal 24-hour solar day. As reported in 1977 by three researchers from Stanford University's School of Medicine and Veterans Administration Hospital in Palo Alto, California, J.X. was awake when the world slept and slept when the world was awake. His body chemistry, temperature, alertness, and performance coincided with the position of the moon. As the scientists concluded, "Throughout the . . . study, there was a remarkable coincidence between his sleep and a local low tide."

While some animals, such as *Limulus*, are guided by the moon via the tides, others may ignore the tide and operate in some mysterious, direct synchrony with the moon. One of the most spectacular of these is a small, luminescent marine organism known as the Bermuda fireworm.

The moon over the West Indies was a day shy of its third quarter when, on October 11, 1492, about an hour after sunset, an eerie light was sighted from the poopdeck of one of Christopher Columbus's ships, the *Santa Maria*. In the captain's log, this light was described as resembling a candle flame being raised and lowered. Was it a campfire on shore? A wishful apparition, envisioned by homesick sailors, of lanterns on the wharves of Palos? A beacon at the edge of the world? The crew never did determine the cause. Many years later some historians suggested that the mysterious light was caused by the spawning of the Bermuda fireworm. Indeed, nearly 500 years after Columbus, beneath a similar moon, in the same location, the tiny worms can be observed conducting their remarkable mating ritual. First they surface and begin glowing intensely to attract a mate. Since their eerie lights are visible only in total darkness, the worms must spawn about a half an hour after sunset and just before the moon rises. This way, no fuel is wasted during times of light. This window of darkness locks the worms' activity into a slot three or four days after a full moon—a time that coincides almost perfectly with the entry in Columbus's log.

From a distance, the spawning looks like a reflection of the aurora borealis on the water. Up close, it consists of millions of tiny blinking green lights, like swarming fireflies. Today the fireworms still mass in this area about an hour after sunset, beneath summer and early autumn moons, for a spectacular display that is visible for miles. The seas turn an iridescent green as the females, both larger and brighter than the males, swarm from their shallow-water burrows and swim to the surface with centipedish legs. They shed a trail of luminescence as they swim in circles two inches or more in diameter. They glow for anywhere from three to eight seconds, briefly rest, and repeat the pattern several times a minute.

The males, meanwhile, still on the sandy bottom, are excited by the glow. Soon they join the females. As they surface, several may be drawn to one female. Each gives two or three quick flashes as it approaches, then dances in a wider loop around her continuous glow. The female then begins to widen her nuptial circle to six inches, the male close in tow. In the course of all the activity, the female releases her eggs while the male floods the water with millions of microscopic sperm. Soon after, the lights dim as males and females spiral back to the bottom. Fertilized eggs drift down with them. This new generation will join in the ritual, beneath the precise summer and early autumn moon, for a few days the following year.

One of the earliest descriptions of this remarkable worm comes from biologist T. W. Galloway, who visited Bermuda in 1907:

The male appears first as a delicate glint of light, possibly as much as 10 or 15 feet from the luminous female. They do not swim at the surface, as do the females, but come obliquely up from the deeper water. They dart directly for the center of the luminous circle and they seize the female with remarkable precision, when she is in the acute stage of phosphorescence. If, however, she ceases to be actively phosphorescent before he covers the distance, he is uncertain and apparently ceases swimming, as he certainly ceases being luminous, until she becomes phosphorescent again. When her position becomes defined he quickly seizes her, apparently about the anterior end of the luminous region, and they rotate together in somewhat wider circles, scattering eggs and sperm in the water.

It is still not known precisely how the Bermuda fireworm maintains this tight schedule. It is not tidal influence, because the tides are minimal where the fireworm spawns. It may be directly influenced by light—though the exact mechanism is unknown. Primitive in its wormy ways, it still possesses an extraordinarily complex eye. It has, in fact, a pair of eyes on each side of the head, mounted on maneuverable lobes, that grow larger just before spawning and are specially designed to detect luminescence—one of the very few cases where the color sensitivity of an animal's eye is limited exclusively to the color of its beckoning mate.

The Bermuda fireworm is one of several species of bioluminescent worms. A much larger species of this marine worm, sometimes known as the Samoan palolo worm, lives in honeycomb tunnels around coral reefs in the South Pacific. In October and November, at sunrise during the last quarter of the moon, tail segments a foot long break off mature worms and swim on their own to the surface. The undulating tail segments stir up a froth on the surface, then suddenly explode, mixing eggs and sperm together. Whole villages set out in boats to scoop up the intact tails and bring them ashore in baskets for use in a celebration marked by feasting on fresh-baked palolo.

Many animals tap into natural cycles in order to solve a problem. The Bermuda fireworm, for example, apparently reads the moon's phases to ensure that it spawns between sunset and moonrise—when peak darkness makes its phosphorescent eggs easier to see. The grunion fish, on the other hand, takes advantage of regular tides to enable it to lay eggs high on a beach, where the eggs can develop for two weeks, after which the next peak tide will wash them back to sea. For many creatures that interact with the tide, timing is crucial to successful reproduction. But few animals exhibit such an exquisite sense of time as the grunion.

Every spring, beneath the light or dark of the moon, John

Olguin, the longtime co-director of the Cabrillo Beach Marine Museum in San Pedro, California, takes family groups, school children, and other interested viewers to the beach to witness the lunar courtship of the grunion. He has been doing it since the 1940s and has introduced some two million children to the grunion's courtship. Many recall the experience with the fervor of a religious vision. Olguin tells of the time in 1978, for example, when he and his wife traveled to Fiji: "We took a plane from Los Angeles to Hawaii, then flew to the big island in Fiji. From there we took a plane to a smaller outer island. Then we took a bus as far as the road would go—about eighty-five miles. We climbed into an ox cart and followed this short trail that ended at an isolated resort. I knocked on the door. A woman answered it and said, 'Aren't you the guy from California who raises them guppies?' I had given her a jar of grunion to hold when she was in the fourth grade and she never forgot it.

"Then there was the time me and my wife were on a train, on the Sumatra Express in Kenya, Africa. Four women in their fifties or sixties were eating dinner in the seats across the aisle. The one closest to the window leans over the sandwich and fruits in her friend's lap and asks, 'Are you still giving grunion eggs to teachers?'"

"'You bet I am,' I told her. I asked how she knew, and she said, 'I got some grunion eggs from you a long time ago when I brought my school children to see those fish spawn—the closest thing I ever saw to a miracle!'"

Indeed, if you stand on Cabrillo Beach at peak tides during spring and summer, you can see a mass of silver-sided, six-inch fish that grunt or squeak. The name grunion comes from the Spanish *grunion,* or "grunter." The female actually emits a sort of squeaking during courtship. Tourist brochures that imply they spawn on every beach in southern California have led many spectators to traipse disappointedly along empty stretches of beach in search of the fish. Consequently, over the years the grunion has achieved a legendary, unicornlike reputation. Cooking and eating the grunion, like the palolo worm, has become a virtual rite associated with the moon. Thousands sometimes gather to catch the fish—by law only bare hands can be used—and cook them over an open pit. Several thousand people sometimes gather outside Olguin's museum. He puts dozens of candles out along the beach at the tide mark. Gathered behind the candle line like pious observers of a celestial event, spectators are struck with wonder at the fishes' uncanny talent for catching the crucial tide onto the beach.

Every two weeks high tide peaks along the beaches of southern California. It is called spring tide and occurs near full and new moons. Debris washed up on such a tide must wait two weeks before another another will reach high enough to carry it back to sea. For that very reason, the grunion must catch one of these peak tides if the eggs that it lays are to be granted two weeks to incubate in the warm sand before being washed back to the ocean to hatch into fry. An hour too early or too late, and the eggs may be left too low on the beach. To provide for a safety margin against laying the eggs too high on the beach, the grunion usually ride up just after the tide has peaked—a good way of insuring that the eggs will be beyond the regular tides, but within easy reach of the next spring tide.

The grunions' courtship is elegant in its precision and poignant in its brevity. The males, who usually arrive on the beach first, are soon joined by thousands of females, who ride waves ashore, then slither their tails into the fluid sand as the wave that brought them washes back to sea. The female will wiggle two or three inches into the sand and deposit several thousand eggs. As she writhes to and fro, several males nudge up to her, arc their slender silver bodies, and deposit milt. The milt filters down her body and fertilizes the eggs, while the males head back to the sea. The female may remain ashore for ten minutes before flipping her way back to the surf. Two weeks later the embryos will be returned to the sea by the next spring tide. Now quarter-inch-long fish, still encased in transparent membranes, they pop from their casements as they are swept through the sandy water.

The courtship of the grunion spans just a few nights a year, lasting for several hours on each of those nights. But in this scant time of the spring and summer moons, on relatively few beaches, entire new generations are born. Given the wonder of the spawning of grunion, it is not surprising that people have added some mythological touches. It has been claimed, for example, that the fish come in only on waves that curl to the left. Some observers have modified this slightly and say they ride the ninth, seventh, or twelfth wave into shore. The grunion have also been credited with sending scouts to reconnoiter the beach, presumably by riding ashore at different points and then reporting back to the schooling hordes offshore.

Over the years, scientists have pondered how a male, faced with thousands of other fish, manages to locate a female in the melee. The issue is not so prickly among large, mobile animals, where good vision and readily recognizable differences between the sexes make it easy to pair up without any special talents. But among tiny fish, stranded on the beach, the cue that brings them together must be quite extraordinary. And so it has turned out to be.

The most obvious thing a female does as she burrows is wiggle. To test the effect of wiggling on the males, Jules M. Crane of Cerritos College, in Norwalk, California, had his students wiggle two-foot-long dowels with the same rhythms as the female grunion. When left still, the sticks attracted no males. But when the sticks were wiggled vigorously (about twice a second), males not only moved toward them, but arched their bodies in characteristic S fashion and released sperm. The males may have been attracted by the small squeak the stick produced when it rubbed against the sand, a squeak that mimicked the "grunting" of females. In any case, the wiggling stick experiment pointed toward the possible cues that draw the sexes together. Each animal—horseshoe crab, fireworm, or grunion—has its way.

One of the remarkable adaptations the grunion has made is the flexibility of its hatching schedule. If, by some quirk, the peak tide fails to reach a bunch of two-week-old eggs, some will die. But others will remain in the sand for another two weeks, unhatched and waiting. This is possible because the egg requires a combination of factors to hatch—the last of which is being swept through the sand and roiling surf, which bursts the membrane in which the fry sleeps. Without this, some eggs will lay in waiting for up to six weeks. Olguin demonstrates this hatching method in a unique way for the some 10,000 preschoolers that visit him annually.

"I have a few clear plastic balls about eight inches across, which can be taken apart. I put a grunion inside of it and tell the children, 'This is how the grunion hatch.' Then we roll the balls through the surf and sand, and I say, 'Now they're getting ready to hatch!' Then I pull them apart; the fish flip out and scurry down to the surf. The children go absolutely wild. 'It's a miracle,' they cry."

Once the children get the idea, Olguin takes actual grunion eggs and puts them in a baby food jar with water and sand. The children shake them to mimic eggs being washed across the sand, and are astounded when the pinhead-sized eggs begin bursting, releasing their captives. The fry are then dumped into the sea.

"The eggs can be kept in a little can in your kitchen for up to six weeks," says Olguin. "That's what has the Japanese emperor so intrigued. Dr. Yata Hameda of the Yokosuka Museum has taken my grunion eggs to the Emperor Hirohito. The crown prince has also written and thanked me for the eggs. He loves the things. They've been raising them at the Marineland of Japan for a year. Everybody loves the grunion."

*   *   *

The grunion, Bermuda fireworm, and horseshoe crab all practice external fertilization, one of the first methods of sexual reproduction, and they all rely on the moon or the tides to bring the sexes together with the precision required for fertilization.

Since life evolved in a sea, and seas are greatly influenced by the moon, it is not surprising that the moon still affects animals of the sea and tide. But higher animals, whose ancestors left the water long ago, also continue to show signs of lunar influence. Even among humans, the 29.5-day lunar month may still hold sway. Not only does the length of menstruation (the word comes from the Latin *mensis,* meaning "lunar month") in women between fifteen and forty years of age coincide with it, but the average human gestation period is 266 days, or exactly nine lunar months. There is even evidence today of a slight increase in conceptions at the time of a full moon. Did our ancestors' sexual cycles closely follow the moon's phases with both ovulations and, therefore, conceptions occurring beneath the blue-lit sky of a full moon?

Whatever the moon's lingering influence over human reproduction, its association with romance is almost certainly as much a consequence of biology as of the poetic imagination. The horseshoe crab, the fireworm, and the grunion are clear reminders of how even the most ancient of patterns remain tightly woven into the present fabric of living things. While the earth has changed, many of the basic forces guiding the evolution of life on it—solar light and heat, water, the moon—have not. With our restless psyche and our tendency to emphasize change over the stable traditions of nature, it is easy to forget how the ancient cycles of sun and moon still propel the courtship behavior of many animals every spring. The existence of all life seems to be patterned with binding regularity. The Buddha likened it to an endlessly turning wheel. As another wise man phrased it, "One generation passes and another comes. The sun rises and sets and hurries around again. All rivers run to the sea . . . the water returns to the rivers, and flows to the sea again . . . Nothing is new under the sun."

Or beneath the moon.

3.

The Rise
of Conflict

Woman wants to live her own life; and the man wants to live his;
and each tries to drag the other to the wrong track. One wants to
go north and the other south; and the result is that both have to
go east, though they both hate the east wind.

—George Bernard Shaw,
Pygmalion

IN 1675, when Jesuit priest Anthansius Kircher set out to calculate
how many creatures accompanied Noah on the Ark, he determined
that God's original complement of land animals consisted of 310
species. Today some 100,000 kinds of mollusks, 100,000 species
of protozoa, 8,600 birds, 30,000 species of fish, and a half million
worms are recognized—a tiny fraction of what are estimated to be
ten to thirty million species all together. In a century in which the
moon has been explored and Mars mapped, many thousands of
plants and animals in our midst have yet to be seen or given a
name. In 1975 alone, 786 previously unknown species of flies
were described.

And speciation is but the beginning of diversity, for within every species are myriad individuals. Like snowflakes, no two sexually reproducing organisms are the same. Each differs in appearance, temperament, resistance to disease in thousands of invisible ways. Hidden genetic byroads will guide each to a different future. Compounding the differences is an individual organism's particular experience: no two live the same moment.

If one were to pinpoint the most obvious difference within many species, it would probably be between male and female. Males and females of some species are so different that they have been mistakenly classified as different animals. Only by discovering some male and female ants and wasps in the act of mating can scientists tell that they belong to the same species. Among four families of deep-sea angler fish, which live two thousand feet down in pitch-black ocean depths, the females may be several thousand times larger than the males. Indeed, scientists were surprised to learn that the tiny parasitic appendages often found attached to the females were actually their mates. Once a male locates a female, he sinks his fanglike teeth into her side, permanently fuses to her body, and withers away until virtually nothing but gonads remain—a steady supply of sperm but a poor companion.

When earth was inhabited largely by clones, reproduction was a relatively placid endeavor—no males or females, no need to attract or compete for a mate. But as the sexes slowly evolved and the gap between them widened, the terms of mating agreements grew more complex, and potential for conflict greater. In some cases, male and female became so disparate in body and mind that pairing required long, drawn-out negotiation to overcome intrinsic differences. Like historical enemies forced into a treaty for the sake of survival, the shaky alliance of male and female was undercut by instinctual mistrust. Charles Darwin suggested as early as 1871 that mating was more a bitter truce than willful embrace.

The female, he wrote, chooses "not the male which is most

attractive to her, but the one which is the least distasteful." A century later biologist Robert Trivers went even further when he wrote, "One can, in effect, treat the sexes as if they were different species."

The relationship of sex to conflict has been fertile ground for speculation ever since the abduction of Helen by Paris sparked the Trojan War. Among scientists, this question has been addressed perhaps most forcefully by Edward O. Wilson. "Sex," he writes, "is an antisocial force in evolution. Bonds are formed between individuals in spite of sex and not because of it." In the view of Wilson and many other contemporary biologists, the root of the conflict lies not with male and female, per se, but with the egg and sperm. The reasoning, in part, goes something like this.

The human ovum is 85,000 times larger than the sperm. So is the ovum of a hundred-ton blue whale, the world's largest mammal, and that of the bumblebee bat, which weighs less than a penny and is the world's smallest mammal. Indeed, the ovum of the world's tiniest mammal is probably larger than the sperm of the biggest. In short, sperm are small, eggs comparatively big.

Placid and passive, the ovum is the nurturer and protector. It is large because it must nourish the blastula, the ball of cells that forms after conception. When an ovum is released from the follicles of the ovaries, it is ushered through the fallopian tube to the uterus. In the course of a woman's child-bearing years, as many as 400 will make this journey. Few—a dozen would be an optimistic estimate—will ever become fertilized and eventually see the light of the delivery room. A blue whale might expect to product ten offspring in its lifetime; a sperm whale, six; a giraffe, ten.

The human male releases between 200 million and 300 million sperm with every ejaculation—enough theoretically to father a population as large as that of the United States. This could amount to some twelve trillion children over a lifetime. These intrauterine raiders, propelled by frantically wagging tails, are the most restless of cells. Once freed from the testes they migrate en masse through the uterine darkness. Their mission is to seek and impregnate.

Sperm are plentiful, eggs a rare commodity. Consequently, male and female must employ radically different approaches to achieving their ultimate goal—to leave behind as many of their own descendants as possible. The female must be prudent in her investments. She must choose a male who will stir them to life with a fine and fit sperm, and who will then remain around if necessary to protect the offspring. One way she can assure herself of a fit male is to engage suitors in a rigorous elimination heat, or courtship routine, to winnow out the weak. Though personal tastes may play a role here, one rule seems to be inviolable: "Only the brave shall win the fair."

The vigorous courtship of the European smooth newt, for example, rigorously tests the strength, perseverance, and skill of the male. As the female sits on the bottom of a pond and the male dives to meet her, he must perform a series of exhausting tail movements before the female will mate. If he can't hold his breath long enough to do them, he must resurface and risk losing her altogether. Forcing the male into a long courtship is one way of winnowing out the unfit—perhaps those with inadequate lungs or energy reserves. Presumably, whether by virtue of size, strength, or cleverness, winning traits contribute to survival. In the language of biologists, such traits are "adaptive."

Males, on the other hand, need not always evaluate females so carefully since their sperm represents a minimal investment. The male can afford to be indiscriminate—may even profit from it. The more females he impregnates, the closer he approaches to his maximum reproductive potential—that is, leaving behind as many offspring as possible. Given the availability of sperm, males are limited only by the number of females they can impregnate. But the female, because of her limited supply of ova, must take the opposite approach. For females, cautious avoidance, or coyness, is the best strategy. Unlike the male, who invests a little energy in many offspring, the female pours great energy into a few offspring. It is often her best route toward her own reproductive potential. Other considerations aside, males may tend toward many mates of dubious standing, while females tend toward a few good ones.

Conflict occurs as each tries to implement his or her own tactics at the other's expense. The male wants to go out rather than sit on his nearly limitless reproductive potential; the female wants him to remain around to protect her limited offspring. The dilemma, in human terms, was summed up by a distraught young mother who wrote to a newspaper advice column: "Dear Beth: I'm sixteen and had a baby last month. The night Jimmy and I went all the way he said he loved me. I figured he would still love me and help me support my beautiful baby girl. He left town."

In short, males and females, because of intrinsic differences between egg and sperm, have adopted vastly different reproductive strategies. From this egg-versus-sperm point of view, females and males themselves are not the principal combatants in the battle of the sexes, but are only proxies in a war being waged by gametes. Dorothy Parker captured the conflict in her poem, "A General Review of the Sex Situation":

> Woman wants monogamy;
> Man delights in novelty,
> Love is woman's moon and sun;
> Man has other forms of fun.
> Woman lives but in her lord;
> Count to ten, and man is bored.
> With this the gist and sum of it
> What earthly good can come of it?

A widely touted case of sexual conflict, observed almost exclusively among captive specimens, is that of the praying mantis, in which the female have been known to cannibalize the male during courtship or mating. At the mere sight of a female, the male freezes, sometimes for hours on end. Eventually, he moves cautiously toward the female, again freezing if she makes the slightest motion. Once within about a body length of her, deliberation gives way to action: he quickly mounts her, clasps her sides with his forelegs, and begins to beat her about the head with his antennae as his abdomen rocks up and down. It may take a half hour before sperm is actually transferred to the female, an unspeakably hazardous time during which the female may twist around and tear off the male's head, then leisurely eat it. Relieved of his brain and its inhibitory center, the male will begin to mate with abandon, rocking his body with a fervor impossible while the brain is still intact.

If the female spots the male first, she may ambush him and eat his head before he has even mounted her. Nevertheless, the preprogrammed movement of his legs will carry his body in a circular path until it rests against her. The headless, thoroughly uninhibited insect then climbs onto the female's back and copulates. Not only may the loss of the male's head greatly increase his chances of impregnating the female by relieving him of his inhibitions, it may increase the survival rate of the offspring by providing the female with a good meal—especially in circumstances during which she has been deprived of food. One way to look at this extreme reproductive adaptation—if, indeed, it is one—is as the ultimate female blow against the male's inclination to unfaithfulness.

While this unusual behavior has rarely been observed in wild praying mantises, female spiders sometimes naturally attack and kill their suitors. In some species, prospective males must signal their peaceful intent by strumming the outer edge of the web, as if serenading the female into compliance.

The fact that an animal's armaments often play a major role in courtship and mating also suggests a close link between sex and violence. For example, the spiral ivory spear of the narwhale, horns and tusks, the teeth of elephant seals, and the antlers of deer can all become deadly weapons during courtship. In some cases, the link between armaments and sexual behavior is explicit. In his classic work, *A Herd of Red Deer,* English naturalist Frank Fraser Darling described a stag masturbating by drawing his sensitive antlers through shubbery. "The sexual activity of the stag is so intense that he has not sufficient hinds to satisfy his lust. . . . He may masturbate several times during the day . . . even when he has had a harem of hinds." After lowering his head and gently drawing the antler tips back and forth through the underbrush, "erection and extrusion of the penis from the sheath follow in five to seven seconds," Darling reported. "There is no oscillating movement of the pelvis. Ejaculation follows about five seconds after the penis is erected, so that the whole act takes ten to fifteen seconds. These antlers, used now so delicately, may within a few minutes be used with all the body's force behind them to clash with the antlers of another stag."

Antlers are a dramatic adaptation to stiff competition between males. Some early scientific observers once thought the antlers were made of wood; they are actually made of keratin, the protein that forms fingernails. Unlike horns, which are permanent, antlers are shed annually. They grow faster than any other bone—sometimes a half inch a day. At first they are covered with "velvet," which contains a rich supply of blood vessels that deposit the calcium and other minerals that gradually harden into bone. As rutting season approaches, the velvet constricts, breaks into tatters, and falls off. Some remaining vessels may stain the white bone underneath a bright red, but as they dry the antlers turn dark brown. The stag may then spar with bushes in preparation for actual competition.

Dutch naturalist Niko Tinbergen believed that most fighting among animals was sexually related. Among some reptiles, sexual rivalry is the only source of fighting. Violent or deadly combat has been observed in insects, muskrats, California ground squirrels, elephants, tigers, hippopotami, musk oxen, grizzly bears, and langurs, to name a few. Zoologist Dian Fossey, who studied the mountain gorilla on the remote slopes of the Virunga volcanoes in Rwanda for almost two decades, reported that during 11,000 hours of observation of less than 300 animals, she witnessed three cases of adult males killing infants and suspected the killing of three others. Infanticide, not uncommon among many animals, including lions, gorillas, and langurs, not to mention rodents, birds, fish, humans, and numerous invertebrates, is often the work of a male who murders his predecessor's offspring, thereby increasing his own genetic stake. Based on her work with langurs in India in the early 1970s, zoologist Sarah Blaffer Hrdy tells the story of the matronly Itch, who with the help of her female friends, old Sol and one-armed Pawless, bravely fended off the repeated assaults of Mug, a male who had moved in on the old male, Shifty Leftless, and was cruelly lording it over the females, attempting to kill their infants and sire his own.

Early one August afternoon, Itch's hillside troop was feeding quietly in the trees, when Mug suddenly advertised himself with a vile grunting sound and climbed atop the roof of a nearby school building. "Mug strikes a sentinel's pose and stares off into the distance," Hrdy writes. "Then, abruptly, the stocky gray form descends from his rooftop perch, charges directly at Itch, and grabs at the infant clinging to her belly. Itch whirls to face Mug, plants her front paws as she lunges at him, grimaces, and bares her teeth." Fortunately, the other females mount a swift counterattack, chasing Mug up a tree, and "a trembling Itch is left alone to hold her infant Scratch, who is spattered with flecks of blood."

After relentless stalking, Mug finally is granted the opportunity—partly through Itch's own carelessness—to seize the infant: "Maternal negligence was almost surely at issue in the second attack . . . when Itch let Scratch fall out of a jacaranda tree. When the infant fell, [Mug] raced to it, reaching it just split seconds before Sol and Pawless. Only by a fierce assault were the females able to wrest the infant from his attacker."

In September, the infant was again attacked by Mug, this time being gravely wounded. A large gash was opened in its thigh and tooth marks perforated its head. Miraculously, the infant survived, only to disappear a few months later—undoubtedly at the hands of Mug.

Mug's assaults were the continuation of a story of dethronement, for when Mug's predecessor, Shifty Leftless, had taken control of the troop several months before, he had exterminated his predecessor's infants and reimpregnated the females. Male and female act not only at each other's expense: the ultimate cost of such selfishness may be paid by the species. Evolution, far from serving to prevent extinction of a species, as traditional zoology has taught, may serve first and foremost the interest of whatever individuals manage to leave behind the most offspring—even if it means killing the progeny of others. "For the good of the species" is a noble but erroneous concept. "Selfish" behavior, in fact, may actually contribute to the high historical rate of extinction of species.

Male-female antagonism is but one aspect of sexual conflict. There is also male competition, which Darwin described as males trying to "conquer other males in battle." Rivalry among elephant seals is among the most brutal violent sexually related behavior. Every spring along certain California beaches, bulls engage in bloody competition for female seals. The fight begins as a gruff shouting match with two males exchanging deep-throated roars. If one doesn't retreat, then the shouting match escalates into combat. Weighing up to two tons each—about twice the size of a cow—they slam their bludgeonlike noses into each other while trying to sink their large teeth into the neck of their opponent. Newborns are the most frequent victims as males throw their weight around, and the beaches resound with the shrill cries of crushed infants. Nearly half of the pups' deaths in a single season are caused by battling males.

Of course, sexual violence is no stranger to human society. Some biologists speculate that those primitive men who exhibited a willingness to fight gained an edge in survival, thereby genetically passing this asset on to their descendants and eventually giving rise to a race of warriors with large anterior pituitary glands—the chemical center of aggression.

If sex is the root of so much conflict among animals, are we left with the unsettling conclusion that the most peaceful societies are unisexual or those that minimize sexual differences? Interestingly, the earthly society that seems to have reached the acme of efficiency and cooperation is, indeed, run by and populated almost exclusively by females: the honeybee hive. The queen and workers are all females, while the role of the male drone seems to be limited to a nuptial flight in the spring. After carrying out this indispensable task, drones are ignored and finally cast out of the hive.

Male and female are intrinsic ingredients in a recipe for combat. So it is ironic that we associate courtship with the apparent tranquility of spring, which, in truth, is a time of heightened conflict. Spring is also the season when life's astounding diversity comes clearly into view—a richness that owes much of its existence to sex. And to which the world owes much of its woe.

4.

Spring

ON MARCH 20, 1986, daylight first struck the bronze statue of John James Audubon, presiding from its granite pedestal over the live oaks and hibiscus of New Orleans's Audubon Park, at 6:05 A.M. The sun set at 6:12 P.M., giving this river metropolis a balmy twelve hours and seven minutes of daylight on the first day of spring. Along the outlying bayous and marshlands where the artist and naturalist searched for subjects to feature in *Birds of America*, bucks and does have mated and are preparing to give birth. Mockingbirds have laid their clutch of three to five spotted, greenish-hued eggs. Surrounding wetlands are dotted with snowy egrets, their gossamer primary tips splayed by the slightest breeze. The grass heads grow, ripen, and explode with seed, while blossoms on the blackberry thickets disintegrate in a breath of wind, leaving behind tiny variegated kernels that will swell into a rich if short-lived source of food for many animals.

Spring is a season of transient opportunity. Lengthening days in the northern hemisphere, caused by the north pole tilting closer to the sun, allow time for locating a mate, for courting, and for eventually raising young, while rising temperatures free animals from the time-consuming task of keeping warm. June 22, the summer solstice, is the longest day of the year. Summer has arrived.

Courtship in the Animal Kingdom

The tepid days give way to hot afternoons. A rasping chorus of katydids covers the land, while the calls of cicadas throb like currents surging through high voltage wires. Days begin to grow shorter. By the end of summer, as night temperatures fall, a layer of humidity will bed down between the Mississippi River and the stars. Come late autumn, but for an hour's backward shift in Daylight Saving Time, the statue of John James Audubon would not see sunrise until nearly 7 A.M.

No matter how short or subtle, spring comes to virtually every land—although spring, of course, is not the only time animals mate. The Arctic spring and summer pass in a brief aura of light. Even the lush, seemingly inexhaustible tropical rain forests of the equatorial regions must pause at some time, shed leaves, and prepare for new growth. The deep, lightless realms of the ocean may be one of the few places on earth where the seasons play no role in the behavior of animals.

Elsewhere, creatures have evolved within the rhythm of night and day for so long that the balance profoundly determines the agenda of their lives, especially in reproductive activities. By artificially alternating the number of hours of "daylight" a redwinged blackbird receives, it can be made to build a nest. Other photoperiods, or time slots of light, send birds on their migrations and determine the time of molt. Subject a greenfinch to an artificial seven-hour "day" followed by a seventeen-hour "night," and its testes will begin producing new tissue, but give the bird an additional seven hours of daylight, and it will actually begin producing sperm.

Photoperiods trigger similar transformations in the female: the ovaries begin to swell in preparation for releasing the season's supply of ova. A female, who only weeks before might have taken flight at the sight of a male, may now actively solicit him. In species in which the female initiates courtship, it may be her aggressive tendencies that are carried on the rising tide of hormones.

49

During breeding season, female red-winged blackbirds in Dutchess County, New York, turn aggressively on one another as they fight for the allegiance of a siring male. Remote as the association may seem, many of these physical and behavioral transformations are tied, in part, to increasing day length.

The sun orchestrates many facets of animal behavior, not least of all courtship. Among other things, it provides the cue that helps males and females synchronize reproductive cycles so that sperm and ova are produced simultaneously. Not only do specific day lengths spur the blooming of both male and female gonads in the spring, but reliance on the sun indirectly leads to a sort of communal birth among the species, which coincides with abundant food—in itself a product of increasing day length. Sheep in the southern hemisphere breed at a certain time of the year, while those in the northern hemisphere breed at another, each in accordance with the best local conditions.

What is the connection between sex and sunlight? Part of the answer lies in the complicated realm of endocrinology—the study of glands that produce hormones. Among the major components of the endocrine system are the ovaries and testes and the pituitary, pineal, and adrenal glands. Indeed, the brain itself secretes some forty-five separate hormones. Altogether, the human endocrine system may produce 200 hormones that regulate everything from body growth to sex drive. Not surprisingly, hormones also form a major link in relaying the message to the reproductive organs that spring has arrived.

Located near the center of the mammalian brain, between the hemispheres of the cerebellum, is a small structure known as the pineal gland. Descartes thought it was the seat of the soul. In fact, it is the clockkeeper for seasonal breeding in many animals. Cutting the nerve to the pineal gland prevents some animal's coats from changing color in winter; if a sheep's pineal gland is severed, it will breed out of season, creating what one wry Scottish observer

50

called "a ram for all seasons." The Syrian hamster requires some twelve and a half hours of daylight to keep its gonads active, and the short days of mid-autumn cause the organs to atrophy as the animals go into hibernation. But if the pineal gland is removed, Syrian hamsters remain sexually active year round. By keeping tabs on day length, the pineal gland helps to assure that females come into estrus and males into mating condition at the optimal time. How the gland monitors day length and then translates this raw data into information vital to courtship is a marvel of chemistry, engineering, and evolution itself.

In many mammals, light is fed to the pineal gland by way of the optic nerve. In most birds, however, it filters directly through the thin skull. In the absence of light—the short days of winter—the pineal gland produces a chemical called melatonin. This substance blocks the nearby pituitary gland from secreting luteinizing hormone (LH), which is vital to reproduction. LH is to the testes and ovaries what phosphorous is to a strike-anywhere match: without it, they are inert. When the long nights of winter cause the pineal gland to secrete the pituitary-stifling melatonin, the testes and gonads remain in a state of relative dormancy. As the days lengthen, the pineal gland reduces production of melatonin, the blockade is lifted, and the pituitary gland responds by sending LH to the testes and ovaries. The testes, in turn, secrete testosterone and other androgens, or male sex hormones, which spur a male's aggressiveness and mounting behavior. Testosterone is one of the major chemical players behind the rise of male courtship in the spring.

In females, LH from the pituitary causes the ovaries to secrete estrogen, which, in some animals, influences female receptivity. With the arrival of LH from the pituitary, a fresh supply of blood floods the ovaries, preparing them for the release of eggs. LH, along with other hormones, later coaxes the ovarian follicles to release the eggs into the fallopian tubes. Yet another hormone produced by the ovaries, progesterone, regulates the passage of the ovum through the tubes, while a complex array of other hormones variously influence both male and female reproductive systems.

While the chemical connection between increasing day length and fertility is a relatively new discovery, more than 2,000 years ago Aristotle must have suspected that the ovaries influenced sexual behavior when he wrote, "the ovaries of sows are excised with a view to quenching their sexual appetites." Far from only producing an egg, the ovaries secrete several hormones—estrogen most active among them—that help support the entire female reproductive apparatus, including mammary glands and sexual drive. Implanted in the hypothalamus of spayed cats, estrogen can, in the words of British researcher B. A. Cross, turn "an aggressively frigid female into a raging nymphomaniac." Generally speaking, androgens stimulate males toward aggression, while estrogens promote receptivity in females. Administered to a male, on the other hand, estrogen can make him *less* aggressive. In one experiment, a red grouse implanted with estrogen lost his hen and his territory. Aggressiveness and receptivity, however, should not be confused with active and passive. The female is just as active in courtship as the male, though often in a more subtle, if not seemingly more dignified, manner.

Estrogens are responsible for a complex and bewildering array of female behaviors. Occasionally the connection between hormones and female aggression is clear-cut: estrogen administered to a female when she is caring for her young will often provoke her to attack. The female reindeer, for example, is passive during most of her life, but she becomes highly aggressive just before and after giving birth. While it is generally conceded that males have largely cornered the aggression market, there are notable exceptions. Female rhesus monkeys usually lead intergroup attacks, female house mice attack more often than males, and when Gelada baboon troops encounter each other, females usually go to the defense.

While the connection between hormones and female aggression is poorly understood, the powerful role testosterone plays in male aggression is well documented. As early as 1400 B.C., Susruta of India recommended that Hindus eat testicles to cure impotence. In 1889 the French physiologist Charles Édouard Brown-Séquard may have been acting out of a similar belief when he removed the testicles of dogs and guinea pigs, combined them in a saline solution, and injected the mixture under his skin. He proclaimed a "remarkable return of my physical endurance." He died, perhaps coincidentally, about a month later. One scientist relates the story of two weak, nonterritorial cocks, who were implanted with androgen and soon regained their vigor, drove off territorial cocks, and established new territories. One of them eventually even won a hen—a spectacular turnabout for a bird that, without androgen, would probably have died.

The potency of testosterone injections or implantations has been documented in a long series of experiments. Among the results: hens have gone from the bottom of the pecking order all the way to the top. A castrated hen so treated will take on the appearance of a cock—with spurs, combs, and wattles—and will even crow and act like a male. Capons have been transformed into roosters that crowed, battled, and chased hens. Administered testosterone, a chick will crow when only three days old. Testosterone naturally spurs the growth of a stag's antlers in the spring; if a doe is injected with the substance, she will grow antlers too. If administered testosterone, female canaries, male and female doves, male chaffinches, Japanese quail, and even female swordfish will all rise in social rank. Fights will erupt among turkey chicks. Ring doves will enlarge their territories.

While the hormone heightens aggressiveness, aggression may

also produce the hormone by causing the brain to signal the pituitary gland to release LH, which in turn stimulates the production of more testosterone. If aggression does, indeed, increase the production of testosterone, violence may, in the truest biological sense, beget violence. Circumstantial evidence for this hypothesis comes from a 1980 experiment in which it was found that after resounding victories over opponents in tennis matches, the winners' testosterone levels rose sharply, while the losers' fell dramatically.

Hormones are a single thread in a complex, still largely unexplained pattern of animal courtship behavior. In general, one can envision them as a sort of chemical Rube Goldberg device—one thing triggers another until a seemingly unrelated chain reaction is tripped into motion, at the end of which an ovum is dropped from the ovaries and sperm released from the testes. Endocrinologists will require many more decades to piece together the full picture.

The story of courtship behavior is, of course, far more complex than just the arrival of longer days in the spring. In fact, *decreasing* fall light plays a role in triggering breeding in sheep, trout, and many other so-called "short-day" breeders. In tropical and equatorial regions, where the length of day hardly varies throughout the year, seasonal rhythm is often marked by changes in climate. The onset of the rainy season, for example, provides the breeding cue for weaverbirds and some cranes. Not surprisingly, many animals in these regions have no pineal gland at all—the anteater, sloth, armadillo, and crocodile, to name a few. Others, such as the rhinoceros and elephant, have very small glands. But as one moves north or south of the equator, where seasons cause dramatic shifts in day length, the pineal glands of resident animals, such as the elephant seal and sea lion, grow increasingly large.

In addition, there are many nonseasonal breeders, which breed year round. There are also the "reflex ovulators" such as cats—

and, some have claimed, occasionally human beings—for whom the very act of copulation triggers release of the egg. Coitus alone may stimulate the central nervous system, causing production of LH, which in turn prompts ovulation.

Might there be some vestigial link between the seasons and human sexual behavior? In 1983 two scientists from Atlanta's Emory University demonstrated that both assault and rape peaked in certain locations between July 7 and September 8. One's first inclination is to attribute this to higher temperatures—the so-called "thermatic law of crime." In fact, statistical analysis eliminated temperature as the determining variable, leading the scientists to hint that, just as many animals experience a dramatic rise in aggressive behavior during rut, human aggression may also be attuned to seasonal cues. Furthermore, since general assaults and rapes peaked at the same time, the researchers speculated that sexual aggression is a facet of general assaultive behavior—clearly the case in many animals. The point here, as in practically all parallels between humans and animals, is not that such theories grant us license to act like animals or that they in any way justify our actions when we do, but that such vestigial holdovers, if indeed that is what they are, offer a fascinating glimpse into the evolutionary roots of our own behavior.

Despite the fact that humans long ago ceased living at the total mercy of the seasons and began inhabiting worlds of artificial light and warmth, our bodies still respond to the arrival of spring in subtle ways, the true extent of which we are only dimly aware. Is there, for example, a hidden biological basis for the tradition of the June wedding in Western culture, or for the Maypole celebration or other spring fertility rites? Although we think of ourselves as independent, scientifically-minded observers of the movements of the earth and sun, we also participate biologically in what, to many animals, is the nonnegotiable season of courtship and mating.

5.

Toward a
Peaceable Kingdom

Animals in nature, contrary to the suspicions of cynics or the hopes of idealists, are neither intrinsically vicious nor altruistic. Competition and cooperation are both nature's ways.

—Stephen Jay Gould,
Illuminations: A Bestiary

IN JUNE the European chaffinch, a bird of multicolored pastels, begins to sing at about thirty minutes before dawn. The cock has a slate blue head and nape, with a cranberry back that fades to olive at the rump, then to gray at the tail. Its belly is pinkish brown and its crest chestnut. The female's colors are considerably more subdued. On their cross-continental migrations chaffinches have been clocked flying at forty-four kilometers an hour, at night, through the Swiss Alps.

The chaffinch sings in various dialects, depending on where it lives. Its song at rest is a gleeful, rattling *pink-pink-pink*, while its flight song is a gentler *tsup-tsup-tsup*. If it spots a perched hawk, it

56

sends out the alarm *chink-chink-chink;* a flying hawk elicits a *wheet-wheet-wheet. Chink* if by land, *wheet* if by air.

Like many animals, the chaffinch finds itself in a bind when it comes to mating. It lives life in a delicate balance between fleeing from and attacking those that come too close. In a world of potentially dangerous strangers, the impulses of fight and flight must be kept in check. Mating intensifies this dilemma. To many animals, physical contact defies the basic instinct that has helped them to survive. Even some of the most gregarious have a threshold over which strangers may not pass. Swallows demand 15 centimeters; the black-headed gull, 30; the greater flamingo, a comfortable 60; while the sandhill crane will settle for no less than a spacious 175. Many solitary mammals require much more room. Even humans have a fear of physical intimacy. As James Thurber and E. B. White noted in their tongue-in-cheek book *Is Sex Necessary?*, "Many men have told me that they would not object to sex were it not for its contactual aspects." So it is with many other animals who, without the reassuring gestures of courtship, might never mate at all.

What does an animal do when, swept up by the urge to mate, it is drawn to the thing it fears? The outcome depends on the relative strength of its conflicting drives. If fear wins out, it flees. If the impulse to attack is stronger, it attacks. Not until the urge to mate overcomes both of these can copulation occur. Courtship then, is a three-way negotiation among the instinct to flee, fight, and mate.

In spring, when the male and female chaffinch come together, the male initiates courtship with a posture in which he raises his wings slightly while dropping his primaries—the principal feathers of his wings. Tail feathers are raised, spread, and sometimes moved up and down. Occasionally his wings shiver. He lowers his head and draws down his neck so his crown is nearly level with his back, exposing a white shoulder "flash." The feathers on his throat, belly, and breast may be slightly fluffed, those on the crown and back drawn sleekly along the body. Sparked by the lengthening days and rising temperature, his testes have been producing testosterone, and he has been advertising his virility with a *tsit-pink-tcheek*. Often he edges close to the female and drops into a rakish "lopsided posture," bending a leg and cocking his body to one side, like a tap dancer preparing to exit the stage. He may walk about in a peculiar series of small steps, pause, then take a few more. He usually maintains a zigzag course, alternately turning around to model one side, then the other. He may remain silent or emit a quick *tsit,* a *chirr,* or even a prolonged *chirrrrrrup,* which sometimes streams off into a rattle. He launches this song with his head held low, crooning like an impassioned balladeer, then throws up his head as the song climaxes. If the female remains unimpressed, the male may vigorously wipe his bill back and forth across a branch in apparent frustration.

Desite the vigor of the male's courtship, he is actually in a state of some trepidation. In fact, in the early stages, when fear still outweighs ardor, he seems so insecure that any movement toward him by the female sends him fleeing. Sometimes she flies toward him and usurps his perch. The male's gesture of subordination seems to appease the female. How close the male approaches her afterward depends, once again, on the relative strength of his temptation to flee over his desire to mate.

The female, meanwhile, has also been engaged in her own private battle. She initially signals her ardor with a timid *tsit-tsit*— from one every few minutes to one every few seconds. Her solicitations usually begin shortly before nest building, climaxing when the construction is underway. Like the male, she drops the primaries of her wings, elevates her tail slightly, and spreads her feathers. Her crest raised, she shivers. The bolder her own posturing grows and the more she is drawn toward her mate, the more fearful she, too, becomes. Like the male, she is locked in a struggle of deciding whether to fight or flee, at the same time she is drawn toward him. Finally, as she prepares to mate, she lowers her breast nearly to the ground, raises her girdle, and bends her legs slightly. She draws back her neck until her bill points sharply upward. She extends her wings. Her crown and upper body feathers are smooth, while those on the breast and belly, especially around her cloaca, become parted. The tail is held almost vertically as she crouches, and she calls out in a rapid *tsit-tsit-tsit.*

The male responds with an acrobatic "upright wings-dropped posture." He stands erect and unyielding as he droops his wings. The more intense his courtship, the more pronounced his fear. He approaches hesitantly, his zigzag course reflecting his indecision. He bends his body to one side, then to the other. If the female meets his advances with an open beak, he expresses his urge to flee with a throaty rattle. At other times, he approaches in song.

As in most courtship, male and female play off each other in a conversation of motions, so to speak. The female's display and calling become more intense, while the male takes flight and hovers a few inches above her, then settles unsteadily on her back, the precarious balancing act maintained by a frantic flapping of both birds' wings. His tail is shifted to one side to enable their cloacas to touch. During copulation he may peck aggressively at her crown, while both launch into feverish song.

The chaffinches' behavior reflects the bewildering array of conflicting drives birds (and other animals) often possess. In many species, an attack is a common response of males when first meeting a female during the breeding season—the chaffinch, snow bunting, the great tit, cichlid fish, and three-spined stickleback, to mention a few. An important dimension of courtship is the constantly changing balance of power between male and female. In winter, for example, the male chaffinch dominates, but the closer the birds move toward copulation, the more dominant the female becomes, until by the height of courtship she has gained a decisive edge. This transition seems to leave both male and female uneasy. The male can be so overcome by fear that sometimes he abruptly halts his display, fluffs out his feathers, and buries his head in his neck feathers in the ultimate gesture of appeasement— the equivalent of sheathing a weapon. Though he may give the female a severe flogging at the outset, his dominance is short-lived. The female, on the other hand, fears the male greatly at first, but as he is forced to mask his aggressive drives in order to draw near, she quickly exploits the lull and goes on the attack, thereby gaining dominance. The greater his desire to mate, the more he must cap his aggression; the closer he comes to the female, the quicker she is to take advantage of the reprieve; the more intense the conflict, the more intense the courtship. From the moment they meet, male and female engage in a tenuous cycle of behavior that unites former antagonists. Courtship is thus the process of narrowing divergent, if not contradictory impulses, until at a moment of peaceful equilibrium, copulation occurs.

It seems paradoxical that those mates who fight most intensely may, in the end, form the strongest bonds. The intensity of "friendship" may, in a sense, be proportional to the partners' potential for mayhem. "Among birds," wrote naturalist Konrad Lorenz, "the most aggressive representatives of any group are also the staunchest friends, and the same applies to mammals . . . the symbol of all aggression, the wolf . . . has become 'man's best friend.'" Similarly, huge flocks of birds or schools of fish live in peace until breeding season, when aggression mounts. This surge of energy is shunted into the formation of pair bonds. At the end of summer, hostilities cool and bonds dissolve. Former mates melt once again into great, peaceful aggregations.

A good example of how aggression can be transformed into a pair bond occurs in the cichlid, a freshwater fish native to southern Asia and Africa. Popular with aquarists, common members of this 600-species family—including the Oscar, Jack Dempsey, and firemouth—can often be seen in pet shops, gill-flaring at their reflections in the glass. When a male reaches sexual maturity, he stakes out a territory and drives off other males. On coming of age, an unpaired female approaches. The male usually attacks. The female flees, but punctuates her flight with appeasing gestures that may override the male's territorial instincts and evoke some latent ardor. She returns time after time, until the male grows accustomed to her presence.

This acceptance depends on a few specific actions on the female's part: initially she must show submissiveness by fleeing, and she must always approach his territory along the same route. In time, she becomes sufficiently self-assured to station herself in the middle of his territory and spread her fins defiantly. Unaccustomed to such a fearless and independent female, the male's aggressiveness suddenly flares again. He displays himself broadside, and thwacks her with "tail beats," sending jets of water against her side. With a furious start, he swims toward her, only to veer away at the final moment. His next actions provide a lucid insight into how aggression can be forged into a force that helps bind the pair together. The furious male knifes past his prospective mate and unleashes his fury against an unsuspecting male in the adjacent territory. The hostility evoked by his mate is transformed into a defense of the area in which they will raise their offspring. Naturalist Niko Tinbergen called this process "redirection." A male who is unwilling to drive off the female in the early days of their relationship might well prove to be a cowardly defender, a weak mate, and an incompetent parent. In a sense, the more hostile he is toward the female, the better father he may prove to be. Aggression has become an essential ingredient in creating the pair bond. Hostility, in effect, stokes the fire that welds them together.

A consequence of conflicting drives is often displacement,

which has become an integral element of many courtships. An animal faced with the choice of either doing harm by attacking or wasting an opportunity by fleeing will opt to vent his frustration by taking a third, seemingly irrelevant course, which neither antagonizes nor alienates its mate. This displacement activity buys time without breaking the courtship cycle. The male chaffinch, for example, when faced with an unresponsive female, wipes his bill back and forth on his perch. A male stickleback in a similar situation suddenly swims to his newly built nest and begins fanning it with fresh water, although no eggs have yet been laid. Through the course of many generations, such displacement behaviors are repeated so often that they become ritualized. No longer serving their original purpose, they nevertheless become genetically entrenched.

Lorenz likened the evolution of ritualized gestures to two imaginary American Indian enemies, Spotted Wolf and Piebald Eagle, sitting down together to resolve a dispute of hunting rights on an island in Little Beaver River. Like the potential mates among many animals, the Indians at first "approach each other in a particularly proud, provocative attitude. . . ." After staring at each other for so long, too frightened to utter a word, one finally vents his tension by doing something neutral—picking up a pipe and smoking it. After many generations, the gesture of smoking a pipe has become a powerful sign of peaceful intent. Similarly, certain animal gestures signal nonaggression. They are meant to appease. So evolved—according to Lorenz—the complex choreography that carries an individual from a first meeting, past the hazards of natural antagonism and antisocial impulses, to actual mating.

Birds provide some of the most striking examples of ritualization. It was the black-headed gull, in fact, that served as one of the earliest models from which courtship was broken down into constituent parts—namely, fight and flight. This small gull is native to southwest Europe, where it frequents coastlines, estuaries, and lakes. Graceful for a gull, it often exhibits a maneuverable, ternlike brilliance in flight. In summer it wears a chocolate-brown hood, and in winter its blacked-tipped, white forewings easily distinguish it from other gulls. Its courtship shows how aggression, carefully managed, can become the essence of coexistence between the sexes. What begins as an aggressive confrontation between male and female eventually concludes with a bout of gentle cooing and mating.

At the start of the breeding season, the gulls, fresh from their spring migration, wheel in from over the sea and settle in the marsh. A gull often returns from its wintering grounds to the same nesting colony where it was born. The males arrive before the females, remaining on the ground for only an hour or two at a time, until reacquainted with the gullery. These early nervous periods are frequently interrupted by what one ornithologist called "silent panics," during which entire flocks suddenly take wing with an almost eerie silence, wheeling over the gullery and slowly drifting groundward again like a constellation, its stars torn asunder, falling to earth. Once a male establishes his territory, he stands quietly or sits down and engages himself in a bout of furious preening. He may take off and cruise above his territory, calling out in an authoritarian *kreeeee-kreeeee-kreeeee*. He lands with a characteristic *kreeooo-kroooo-kroo-kro-kr*. He aggressively defends his territory against any intruder, including females, who, drawn by his calls, often alight nearby. In such encounters both birds confront each other threateningly and adopt a number of interesting postures that quickly neutralize the escalating conflict. One of the most common is the upright posture—a stiff-legged walk, tail nearly parallel to the ground, beak pointed downward at about forty-five degrees, neck puffed out, while elsewhere the plumage becomes marble smooth—a posture often used along the border of another bird's territory. The tendency while crossing into foreign territory is to attack; but once outside, it is to flee. A compro-

63

mise between these two tendencies, the upright posture is also found in the early stages of courtship.

Spreading its tail while stretching its neck forward is the way in which this gull prepares for flight. The thickened neck suggests that the bird is trying to bring its neck forward, while at the same time pulling it back. The downward inclination of the bill is prompted by a desire to peck. In fact, what it often ends up doing is adopting the "upright posture"—a blend of the two contradictory behaviors. Other steps in the courtship dance reflect a similar "compromise" between opposing forces. Often a bird orients its body neither directly toward nor away from its mate, but vacillates, like a compass needle searching for magnetic north in a junkyard. The orientation of the bird's body fluctuates somewhere between full attack and total escape. The greater the urge to escape, the more it turns away from its would-be mate; the greater the urge to attack, the more it faces it. Similarly, the bill reflects the same conflict—the tendency to flight raising it, the urge to fight lowering it—a virtual weathervane of the bird's intent.

The behavior of both the chaffinch and seagull suggest that elaborate courtship dances may actually be aggression in disguise—tension transformed into artful motion. Or as noted ornithologist Edward Armstrong phrased it, such ceremony is the evolved antidote to clumsiness, disorder, and misunderstanding.

The conflict between potential mates is but one front of reproductive conflict. While potential mates vent their aggression during mating, males are engaged in a battle for females. As male and female may limit conflict through courtship, so battling males often use similar appeasement gestures. This is not surprising, since they, like male and female, are also faced with the dilemma of whether to fight or flee.

While male competition not uncommonly leads to death or severe injury, given the intensity of this conflict among many species, it is surprising that they don't kill each other more often. But in a manner in keeping with an individual's own best interests, animals capable of hurting each other most use their deadly assets least: rattlesnakes resort to a so-called combat dance instead of using their deadly venom in the quest for dominance. Raising the first third of their bodies into the air, they strike blows at each other like rapiers until one collapses or surrenders. Rams, even when given the opportunity, rarely strike at each other's vulnerable flanks. Instead, they align themselves as if in a gentlemanly duel, step back a few paces, then crash head-on so that the enormous shock is absorbed by their horns. In all these cases, minimizing the conflict makes sense for the individual, for to abandon gentlemanly restraint would invite an opponent to retaliate with his own maximum response. It is as if nature's etiquette dictates that a male should focus on breaking his opponent's spirit, not his back, so to speak. Press the advantage only until one submits. Anything beyond is a waste of time and energy. Sometimes a simple signal—a roar or hostile posture—is all that is needed to settle the dispute.

The antlers of deer, for example, are usually proportional to the size of the animal. Used in a show of force, they reveal what an opponent is up against, often sending him fleeing without confrontation. It is when two similarly endowed males meet that they must prove superiority through battle. They clash, lock antlers, and engage in a prolonged tug-of-war until one is exhausted. Serious injury is rare, death rarer still, usually the result of permanently locked antlers which dooms both to starvation.

Even elephant seals, whose battles are often bloody and undoubtedly painful, employ signals to prevent conflict from becoming deadly. A male raises his forequarters, swings back his head, and lets his fleshy proboscis dangle in his gaping mouth, while he emits a loud, low-pitched guttural noise. Intimidation at its finest.

But it is only when an opponent presses the attack that actual combat occurs.

In rare instances, as among some lizards, a gladiatorial fight-to-the-death is the rule, but far more common are harmless, ritualized duels. Take the lizard species *Lacerta agilis*, whose combat is the reptilian equivalent of arm wrestling. First they square off, stand side by side, and face opposite directions. Flexing their ribs, they make their bodies high and narrow, creating the illusion of great size. Then each male pivots his head toward the other, and each in turn, in a most gentlemanly, nonchalant manner, offers up the armored back portion of the head to the other, who quickly seizes it in his jaws and viciously shakes it. The weaker animal usually goes first, so that, if the opponent proves immovable, he can abandon the contest before his stronger rival gets a chance at him. If the contestants are equally matched, however, they ritualistically trade blows. One eventually acknowledges defeat by lying flat and slapping the ground with all fours in a gesture of "I give in." He then scurries away. Were one to attack the other anywhere other than the armored occiput, severe injury would undoubtedly occur. In a closely related species, each offers a knee to the other. Gripping it in each other's mouth, they dance around in a sort of wrestling match.

Toads, too, limit fighting between males. In the European common toad, for example, competition is so intense that females are often smothered beneath the onslaught of suitors, yet few males are injured. The mating season begins when males leave their inland winter burrows and begin their short, nighttime migrations toward a pond. En route, they often intercept a similar migration of females. A male jumps on a female's back and slips his arms tightly under her armpits. If a second male also manages to secure a grip on her, a lengthy struggle follows. Lasting up to twelve hours, they are thought to be the longest fights between any animals. The female, meanwhile, continues to struggle toward the

pond where numerous males will join the fray, tugging and pulling or trying to pry off other males. Throughout the competition,
each male evaluates the competition by the sound of his peeping.
Large males have longer vocal chords and therefore a deeper and
more resonant call or croak than small toads. As small deer avoid
those with larger antlers, small toads shy away from deep croakers.
To test the hypothesis that small toads are, indeed, frightened by
the deep croaks, a biologist devised an experiment in which a
small or a large toad was placed on the back of a female. His
calling was stifled by looping a rubber band through his mouth,
like a horse's bit. A medium-sized male was then placed nearby.
When the unmounted toad began attacking, a recording was
played from a loudspeaker. Regardless of the size of the mounted
male, a deep croak from the loudspeaker stopped the attacker,
while a high peek goaded him. Consequently, unequal rivals barely
meet, with the result that few males are injured.

But if in the battle for females the most aggressive males win
mating rights, what prevents a continual escalation of violence
within the species, with each generation growing more aggressive
than the next, until a fight to the death becomes the standard?
Nature seems to have severely limited aggression. The Arabian
oryx, for example, has extremely long horns, but evolution has
turned them backward to such a degree that males must kneel
down and lower their heads between their front legs in order to
stab each other. The design is the equivalent of a safety latch on a
loaded pistol.

It is largely through such evolutionary limitations on aggression and the appeasment gestures of courtship that mating becomes, in effect, a peaceful interlude in a season otherwise
dominated by sexual conflict, not only among competing males,
but between male and female as well.

Admittedly, the temptation to muse about the possible role of
dance in the formation of human pair-bonds is great. Might the

ritualized motion of dance, in certain instances, amount to a constructive venting of the aggression inherent in the formation of human pair-bonds? In any case, the link between sex and aggression in human behavior is obvious. "For [man] the battle impulse and the sex impulse are as closely intertwined as they are for the stag that battles when in heat," wrote Curt Sachs. "The ecstasy of blood and of love flow together in life as well as in the dance, which here, too, uses life itself as a model: an erotic dance is often added to the war dance, or in taking part in a war dance, women become sexually aroused."

The description is reminiscent of observations, reported many years ago by anthropologist Monica Hunter, of the South African Pondo tribe, in which women, like female prairie sharp-tailed grouse, sometimes seem to have encouraged males to fight by accompanying warriors to battle and watching from nearby hills, singing salacious songs and hiking their skirts above their waists to expose themselves. If a younger woman did not oblige, she would be admonished by an elder: "Don't you know that the army is at grips?"

Courtship is not, of course, a panacea for sexual conflict, though it often does provide means of redirecting aggression into the mutually beneficial purpose of reproduction. Insofar as it neutralizes aggression, courtship may indeed be the road to a kind of momentarily peaceable kingdom.

6.

Song

FOR MANY ANIMALS, sound plays a role in courtship that we humans have difficulty appreciating because, by and large, we do not need to travel far to find a mate. To many more peripatetic animals, however, life would be lonely indeed without the aid of sound. If humans had to travel a proportionate distance to find a mate as, say, the Canadian bird *Aedes vexans*, which ventures about fifteen miles from its birthplace, they would have to cover an area of thirty million square miles, or half the land surface area of earth. Such comparisons are of little use scientifically, but they do help to put things in perspective and to show the crucial role of song and other sounds in helping an animal locate a mate.

Far and away the majority of such signals elude the human observer. Consider, for example, the imperceptible vibrations produced by a small insect known as the water strider, which gently transmits its courtship waves over the surface of a pool to attract a female. Or the species of tropical wandering spider that communicates by sending vibrations through the stalks of the banana and agave trees in which it lives. The complex sequence begins when the female deposits a chemical on the underside of the leaf; when the male detects it, he sends out a train of 76-hertz (Hz) vibrations. After every tenth pulse, the female responds with a brief pulse of her own, created by trembling her legs for a fraction of a second at about 50 Hz. Sometimes she supplements the rhythm with a higher-frequency drumbeat created by rapidly knocking her belly against the leaf. Using specialized sense organs, the spiders can communicate in this manner from over a meter away.

Until recently it was believed that praying mantises were deaf. But scientists recently discovered a "cyclopean ear," a long narrow groove underneath the metathorax, the rear body segment to which the back pair of legs is attached. Researchers subjected captive mantises to ultrasonic sounds and found that their brains responded. But when a drop of melted wax covered their "ear," the insect became deaf. It is thought that the mantises may transmit weak ultrasonic signals to each other during courtship, as well as use the "ear" to detect the ultrasonic sounds of marauding bats which prey upon them. They may also emit their own ultrasonic sounds during courtship. Having only one ear makes it difficult to pinpoint where a sound is coming from, but the mantises may compensate by rotating their bodies like a radar dish. They are the only animal known to have only one "ear."

Researchers at Cornell University recently discovered that elephants send out low-frequency vocalizations, inaudible to humans, which are probably used to communicate over both long and short distances. The sounds were first discovered when wildlife biologist Katy Payne was observing the giant animals at the Washington Park Zoo in Portland, Oregon, and felt the air about her throbbing with sensations like "the slight shock waves one can feel from far-off thunder." As the sounds are made in the elephant's throat, a spot on his forehead flutters. (This spot is actually a skin-covered cavity where the nasal passage enters the skull from the trunk.) These elephant sounds are an important addition to an already impressive wildlife choir that includes the ultrasonic squeals of bats, the howls of wolves, the soprano voices of porpoises, and the haunting "tenor-to-bass" songs of humpback whales. The elephant, however, is the first land mammal known to have obtained a place in the choir by using infrasonic calls.

Some species of eels and fish use electrical current to communicate with a potential mate. Electricity has the advantage of being a private channel. Unlike the croaking of frogs or the flashing of fireflies, a weak current knifing through water cannot easily be pinpointed by a predator. Some electric fish send out electrical currents identifiable only to members of the same species. Before sexual maturity, the signals of the male and female *Sternopygus macrurus* are the same, but then they diverge, with males assuming a low frequency of 50 to 90 Hz, while the females have a higher "voice" in the range of 100 to 150 Hz. In 1971 a researcher in Guyana discovered a male's tone to be exactly one octave below the female's just before spawning. But once they paired, the frequency of either one could be raised or lowered to imitate the other, thereby facilitating mate recognition.

Many animals—birds, primates, and frogs among them—engage in sophisticated duetting, each partner making sounds at precisely alternating intervals. Some of the most spectacular courtship duets are produced by frogs. "The wailing of thousands of spadefoot toads . . . in a Florida roadside ditch, in the pitch-black darkness of a hot summer night, brings to mind the lower levels of the Inferno," wrote E. O. Wilson. Some frogs, in fact, gather into trios or even quartets to increase their chances of luring a mate from the competition. In the classic Greek comedy *The Frogs*, Aristophanes provided one of the earliest phonetic renderings of animal sounds in literature, when Dionysus, sore from rowing across a swamp with Charon, is taunted by them:

We are the swamp-children
Greeny and tiny,
Fluting our voices
As all in time we

71

Sing our koáx, koáx
Koáx Koáx Koáx

The notion that song plays a role in the courtship of many animals—especially birds—is not new. In 1702 Austrian ornithologist Baron von Pernau poignantly discussed the role of song in helping mates keep track of each other:

> The Wood-Lark eagerly follows the attraction call, in contrast to the Skylark that does not care about it; the reason for that difference probably is that God's inexpressible wisdom . . . did not implant in Skylarks that method of attracting each other, because they can see their companions on the flat field and can find them without such help, whereas the Wood-Larks, when flying among bushes and over completely wild ground would often lose each other if they did not utter the attraction call constantly.

The repertoire of bird sounds is vast, though not all their sounds are vocal. Vultures and storks have no voice. The ruffled grouse creates a loud drumming with its wings. The turkey cock creates a clicking sound by catching its foot on the quills of its dropped wing feathers. The male coot "sings" by standing on its nest or a weed raft and slapping its foot loudly against the water, while the kiwi stomps the ground.

Many vocalizations are considered "calls" rather than songs. Calls are short utterances that occur a few at a time and are usually associated with nonsexual behavior such as migration, feeding, or flocking. Like the chaffinch, some birds use one call to warn of a pending attack from the air, another for one from the ground. Many small perching songbirds emit thin, high-pitched whistles, squeaks, or other alarm calls, which tend to be very difficult for predators to pinpoint. There are documented accounts of bushtits, when approached by a sharp-shinned hawk, joining in a shrill, quavering two-minute "confusion chorus." Flooded with noise from so many sources, the hawk cannot locate any one bird. When approached by a cat, small birds often respond with a contagious series of short notes of widely varying frequencies. Some sounds may also even indicate the degree of threat. If frightened by a person or a cat, a bushtit might respond with a high-pitched *tik-tik-tik* instead of its customary low-pitched and alarmed *tschunk*. One of the most spectacular cases of a message-bearing call is that of the African honey guide. Once it locates a wild bee hive it may fly to the nearest human, emitting a series of *churring* notes. Then it flies off and calls again, attempting to lead the human back to the hive. When it finally reaches the destination, it falls silent and waits for the person to break open the hive to get at the honey. The bird then feeds upon what remains.

In contrast to message-specific calls, songs are usually a definite arrangement of notes, often repeated. These are frequently associated with reproduction and are often at least partly learned. Sometimes embellished by creative individuals, songs are infinitely various and rich. A wood lark was once recorded singing more than a hundred melodic outbursts, each consisting of some sixty-eight to eighty notes a second, in five minutes. A tropical manakin was observed to sing for eighty-six percent of the day, or about 6,000 separate times. In 1954 ornithologist L. de Kiriline observed a red-eyed vireo sing 22,197 songs in a single day. Clearly, an activity consuming so much time and effort must be important.

While the ability to fly helps birds to get away with such outlandish advertising, toads, flightless crickets, and a range of land-bound animals also manage to live with its inherent dangers. Among birds, singing has many advantages over other forms of communication: it requires little energy compared to movement; the range of notes, the intensity, and the length of song make it a versatile language for conveying an assortment of messages; it may carry through air, darkness, and impenetrable thickets; finally, it vanishes instantly, leaving no trace for predators to follow.

Given the beauty and apparent spontaneity of the song, one might suspect that birds are moved by the spirit itself. But while one cannot rule out that birds sometimes sing from sheer exuberance, song is tied to hormones, as are most other facets of courtship. Increasing day length not only stimulates production of ova and sperm, but also prompts an outburst of song. With the arrival of winter, sex hormones subside and birds fall silent again. But if injected with a hormone, many birds will burst into song, even in mid-winter. In the 1950s many bird enthusiasts who bought singing canaries from their local pet shops were sorely disappointed when, a month or two later, their birds fell silent. It was later discovered that the "males" were actually testosterone-injected females. When the hormone wore off, so did their song.

In addition to day length, a bird's ardor seems to be strengthened by intensity of light. A bright moon tempts morning-singers to earlier song. The mockingbird will spend the entire night serenading the moon. Traumas may also cause birds to break out in song. After the Chilean earthquake in 1960, it was reported that birds all through the countryside burst out in a loud and discordant chorus.

In fact, song may serve many purposes: it can reveal an individual's sex, serve as a territorial warning, strengthen the pair-bond, encourage ovulation, entice a pair to copulate, mark the changing of the guard at the nest, distinguish friend from foe, or signal one's level of dominance—a sort of vocal muscle-flexing.

One way to learn how song works in practice is to go to the woods, and look and listen. As Yogi Berra once said, "You can observe an awful lot just by watching." Or, you can do what ornithologist Fred Wasserman does: take a recording of the song of the white-throated sparrow to the woods of New Hampshire, duck behind the boughs of a tamarack, turn on the tape player, and watch what happens.

It was mid-July, in the middle of an Eastern seaboard heat wave, when Professor Wasserman, along with eight students from his summer course at Boston University (Animal Behavior 507) and me, left metropolitan Boston for the hour and a half drive north. Near the town of Peterborough, New Hampshire, we turned off the main highway and navigated a narrow road up Pack Monadnock Mountain, soon arriving at an altogether different, chillier climate than the one we had left. As we climbed, hardwood trees gave way to deep groves of conifers, which in turn gave way to a panoramic blue sky. Though it was mid-summer in New England, flocks of birds on an arctic schedule were already migrating south. The plovers, sandpipers, and knots I had witnessed feeding on the eggs of horseshoe crabs along Delaware Bay in June had already been to the tundra and raised their broods, and were now navigating their return flights to Chile and Argentina. For a moment, the never-ending web of time, geography, and reproduction was blindingly apparent.

When we reached the mountaintop, the preserve manager, dressed in a tan New Hampshire Recreation and Parks Department uniform, stepped from a tiny box office and held out a thick roll of tickets. Wasserman rolled down the window of the Honda Prelude.

"We have to pay?" he asked, as if nature could be put on sale.

"You wanna stay?"

"How much?"

"Dollar each," the ranger said.

Wasserman counted out ten dollars. Three cars pulled to one end of the parking lot, and as we got out, a steady wind rustled tops of the spruce, carrying with it the rich evergreen aroma of summer growth. A blue panorama of mountains spread around us, while a cover of cloud, its edges whipped into fine, wispy cirrus, slowly moved overhead. We sat down on a wide, flat rock at the edge of the forest, while the bearded Wasserman climbed a small incline and, standing like Moses, spoke of the song of the white-throated sparrow:

"The males arrive here about the last week in April. This is about two weeks before the first females arrive from wintering grounds further south. The first thing a male does upon arrival is pick out a territory—usually about an acre and a half, with some clearing, often edged with low growth where he can build a nest, and bordered by taller spruce as you see here." Wasserman swept an outstretched arm across creation. "As other males arrive to establish their own areas, they engage in border skirmishes, challenging each other with loud territorial songs, sometimes resorting to actual physical attack. One marks the edge of his territory with song, while the contesting bird launches his own song. After persistent border probing by competing males, territories are agreed upon. So it goes, acre after acre, until entire mountaintops are surveyed and the patchwork plots staked out by song."

The jockeying, he explained, is usually over by the time the first females arrive in early May. They begin visiting the respective males, selecting those with the most suitable territories for nest sites, with an ample supply of insects and other bird-aware assets. The males, meanwhile, court them, and any other passing female, intensely with their song—a couple of high tones followed by a series of lower ones. Once used to drive off a male, the song now attracts females. Once a female selects a male, however, singing quickly tapers off, while the territory shrinks. His singing and the lavish size of his territory have won her, so there seems little point in maintaining either. The female selects a nest site, builds the nest without the male's help, and then, stimulated by his periodic early-morning songs, prepares for copulation. Shortly afterward, she lays four eggs, which hatch within two weeks. By the end of

June a new generation of white-throats takes flight above the ever-greens.

Because the birds nest near the ground, half of the chicks may be eaten by snakes or squirrels, forcing many pairs to begin again, even while others, their reproductive chores completed, prepare to fly south. In mid-July it was the thin, two-tone notes of these late-nesters that rose from the backdrop of trees as Professor Wasserman spoke.

"Hear that?" He cupped a hand behind one ear and turned his head. Then he picked up the black leather case by his feet containing a reel-to-reel tape recorder, and together we traipsed further into the woods. Arriving at a secluded clearing, Wasserman set down the tape recorder and silently directed the students to fan out into cover. He then turned on the recorder and ducked behind a nearby tamarack. The piercing two-tone whistle rose from the speaker. Then again. Almost immediately an irate male slashed down from a nearby spruce, made a reconnaissance flight over the black-cased, rectangular intruder, and took guard in a nearby tree. He swept near the recorder, trying to get a visual fix on the mysterious interloper. Back and forth he arched, his momentary perching followed by his own strident song. At one point he furiously alighted and pecked at the unresponsive speaker before fleeing. When the tape finished, Professor Wasserman stood up, obviously satisfied at having provoked the bird. "You see," he said, "although song may be aesthetically gratifying to humans, it is serious for a bird." He explained that song can repel, attract, even help male and female tune their reproductive cycles so that egg and sperm become available at the same time. Wearing a somewhat disgruntled expression, Wasserman refers to an experiment in which a group of female ringed doves were isolated and a male's song piped into their enclosure. Another group was isolated without music. The ovaries of each group were later removed and their weights compared. Those from the birds with the benefit of song were larger and better developed.

Although Paleolithic bone bird whistles suggest that even early man may have tried to imitate bird sounds, people did not generally appreciate the aesthetic quality of bird song until after the Renaissance. It was Czech composer Anton Dvorak who called birds "the true masters." But had he had the benefit of hearing the song of a much larger and more secretive creature, by whose very oil he may have lit his study, the composer might have reconsidered.

In October 1985, an apparently disoriented humpback whale, later known as Humphrey, headed under San Francisco's Golden Gate Bridge, crossed the bay, and swam up the Sacramento River, where whales don't generally go. For twenty-one days scientists, sorcerers, parapsychologists, military men, politicians, outdoor enthusiasts, fun seekers, and self-proclaimed whale gurus tried to change his course. Acting on the implicit assumption that coercion, not understanding, underlies man's relationship with animals, they vainly tried to *force* Humphrey back down river: they played the sounds of a humpback's mortal enemy, the killer whale; they clanked pipes against the sides of ships; they shouted; they tried to head him off with a flotilla of boats; they even set off explosives in the water. At the same time, many concerned scientists attempted to apply their own knowledge of whale behavior to rescue Humphrey. One of them, Louis Herman, director of the Kewalo Basin Marine Mammal Laboratory in Honolulu and an authority on whale and dolphin behavior, decided to try to lure rather than force Humphrey down river.

Herman's idea was to play sounds of other humpbacks, recorded at an earlier date in the whales' Alaskan feeding grounds. Sure enough, when a speaker was hung in the water and the recordings played, Humphrey's huge head soon broke surface. He moved to within a hundred feet of the boat, and for the next seven hours diligently followed it fifty miles down river. At dusk, as Humphrey and his iron escort neared the San Francisco bridge, the speaker broke. His world suddenly silent, Humphrey refused to cross under the orange girders of the Golden Gate. The following morning, with speakers repaired and water reeling with sound, Humphrey was led beneath the bridge and out to sea.

Lou Herman, along with his associate Joe Mobley, Peter Tyack of the Woods Hole Oceanographic Institution, and several other researchers, has used similar "playback" approaches in attempts to unravel the meaning of humpback song. One of the most haunting and compelling of all animal vocalizations, humpback song may be the longest, slowest, and loudest sound in nature. At one moment it wells with anticipation like a stringed quartet in the Grand Canyon, with wavering violin arias prevailing above the cello until overpowered by a groaning bass. When sung in underwater canyons, the sound reverberates like an organ in a cathedral. Peter Tyack describes the experience of climbing into the water with a singing whale: "Your lungs resonate with the sounds; it is very loud and you feel it throughout your body—you feel it more than hear—it is very eerie to be sitting thirty feet below the water having your body reverberate with the sound of a whale."

The song has proven so complex—so full of innuendos, subtle changes in structure, content, pitch, and intensity—that even after three decades of analysis, scientists have not deciphered its meaning. Not until 1949 was the first whale sound even recorded; it was of the beluga, or sea canary, so called because its whistles and squeaks could sometimes be heard in open air. In 1952 O. W. Schreiber recorded humpback songs for the first time from a U.S. Navy underwater listening post off Kauai, Hawaii. He did not know at the time, however, what produced these eerie sounds from the deep. They were finally identified as humpback sounds by William Schevill of the Woods Hole Oceanographic Institution. In

1968 Roger Payne and Scott McVay discovered that the sounds are produced in repeated patterns, or songs. A long-playing record, *Songs of the Humpback Whale*, subsequently became a best-seller.

Frank Watlington, an acoustic engineer at the Columbia University Geophysical Field Station, made an extensive collection of recordings at the Palisades Sonar Station on St. David's Island, Bermuda. He placed a hydrophone in Castle Harbour, off St. David's, ran a cable to his office, and when the whales sang, turned on his tape recorder. He could then glance out his window and see humpbacks breaching.

Watlington's collection of songs provided enticing material, stored on miles of magnetic tape, that would eventually prove to be a goldmine of scientific information. The sounds were analyzed on spectrographs, machines that rendered visual representations of the songs' frequencies and rhythms. Pioneering scientists, most notably Roger and Katy Payne of the World Wildlife Fund, and Long Term Research Institute, subsequently attempted painstakingly to analyze the songs' structures and, like some ancient, exotic texts, unravel their grammar and meaning. Structural analysis proved some of the songs to be virtual thirty-five minute symphonies, repeated for hours on end. One humpback, recorded in the Caribbean by Howard and Lois Winn of the University of Rhode Island, sang continuously for twenty-two hours before the researchers ran out of tape and packed for home—with the whale still singing.

On first impression, a whale's song seems so complex and full of variation that several whales would seem to be involved. Among the almost infinite variety of sounds, however, a few can be easily recognized: rapid pulses interwoven with sustained tones; short, high-pitched notes that climb the scale; and long, low moaning bass notes. Analyses by the Paynes and their associates have shown the songs to be composed in a remarkably organized fashion. They basically consist of phrases, a sequence of sounds lasting perhaps fifteen seconds. A phrase, for example, might begin with a downward sweeping note, then a faint warble, followed by two similar notes, and capped off with a couple of percussive grunts. Each phrase is usually repeated a number of times. Likewise, a different type of phrase is then sung and repeated. Similar phrases are strung together into a theme. Finally, a song may contain anywhere from two to ten themes. What is so remarkable is that each singer seems to improvise on the number of phrases in each theme but always repeats the themes in the same order.

The humpback has often been lauded for its ability to memorize such long songs. Because songs, as Katy Payne found, undergo complex changes in content and whales continually pick up on one another's songs, the whales must have a considerable capacity to memorize—possible evidence of high intelligence. What is even more remarkable is that, by and large, whales stop singing for about three months a year in their summer feeding grounds. Before the onset of migration to winter breeding areas—such as the Caribbean for the North Atlantic population and near Hawaii for the northern Pacific population—the whales pick up with the same song that ended the previous breeding season. Retrieving a song after three months of "cold storage" is an impressive feat.

But while the new breeding season begins with the same song that ended the previous year's season, the song can change almost beyond recognition in the course of a few breeding seasons. Sometimes the song may more than double in length in a single season. Entire themes may disappear, while new ones are incorporated. In studying songs of whales near Bermuda for more than twenty years, the Paynes found that some songs changed so completely that not a single note was left intact.

As is the case among birds, all whale sounds are not songs. In fact, the large majority of sounds produced by cetaceans—the pulses of killer whales, the clicks of sperm whales, and the vocalizations of finbacks and southern right whales—are just that:

sounds. And while several species are given to song, none comes close to the elaborate humpback's. Unlike the humpback songs, the sounds of some of the other large whales seem to have evolved to travel great distances, and probably serve as contact calls. Just as two hikers separated in a dense wood might whistle to locate one another, whales probably use sounds to track each other when out of sight. Sperm whales, for example, disperse when underwater. But as they surface every half hour or so, they break surface not only at the same time but usually with only tens of meters between them. It is probably their calls that enable them to adjust their spacing and to surface simultaneously. Sound would, indeed, enable whales to communicate over great distances. Impulses traveling thirty-five miles beneath the ocean, probably produced by fin whales, have been heard. In a deep ocean channel, finback sounds may travel thousands of miles. Before propeller-driven ships polluted the oceans with unnatural noises, a whale sound might have traveled nearly four thousand miles through the relatively undisturbed waters—or well across any ocean.

While such sounds probably serve as contact calls, an explanation of the humpback's song is not nearly so simple. For one thing, its high frequency and low volume means that it is best equipped for short range communication. What's more, it's far more complex than would be needed for notifying one whale of another's presence. Since no other whales in the humpback's range sing, it probably doesn't serve as a barrier against humpbacks breeding with other species of whales. Finally, despite the great diversity of sounds within a song, songs probably don't carry specific messages about the singer, since the song changes frequently over time. A "language" that drastically changed from year to year would be obsolete for purposes of communication within a short time. What, then, is the meaning of the humpback's song?

The process of discovering the meaning of humpback song might be likened to someone seeing, for the very first time, a strange creature lumbering slowly behind a picket fence. Only narrow sections of fur and color are discernible between slats. At one time or another each inch of the lumbering animal, which is constantly on the move, reveals itself, yet even its general shape is difficult to deduce. Only by patiently piecing together glimpses into a coherent whole can the mystery be solved. In the case of whale song, scientists are only part of the way there.

With one possible exception, all identified humpback singers have so far proven to be males. They usually don't sing in groups, but generously space themselves across the breeding areas. What's more, apparently receptive females approach singers. This in itself is highly suggestive that song plays an important part in courtship.

One plausible courtship theory proposes that males use song to challenge other males. These "song-as-threat-display" advocates believe that the males use song to establish a dominance hierarchy. The one that establishes his dominance is favored to mate. Other researchers, however, such as Peter Tyack, who has seen females approach these singers, believe that song serves primarily to attract females, not to threaten other males. Of course, song could serve both of these functions simultaneously. In fact, the song could serve many purposes, including marking a floating territory or even serving as a cue by which females synchronize ovulation.

If Tyack is right, then females may have the bigger say in the mating process, with a song's incredible complexity offering the female some measure of evidence upon which to choose a suitable mate. One can imagine many males singing over a wide area and "displaying" to females. Based on their songs, the females can then make a choice among them.

If, indeed, the female bases her choice upon his singing, how might a male's song reflect his fitness? The question is open to speculation. It could reflect such things as how long a whale can hold his breath (since they usually sing one song for each breath).

A few creative researchers have even toyed with the idea that singing ability may reflect how adeptly a male uses air bubbles—in itself an important survival skill. Some humpbacks, for example, ineniously trap schools of fish in bubble nets by standing on their tails on the bottom and releasing a circle of bubbles from their blowholes. This rising bubble net encircles the prey. The whale then opens its mouth and surfaces through the vertical bubble corridor, conveniently drawing the fish into its mouth.

Furthermore, a humpback may use air bubbles in defense—a theory fortified for marine biologist Paul Forestell, Senior Researcher at the Kewalo Laboratory, by a near-mishap he experienced while diving with humpbacks off Hawaii. Equipped only with snorkeling gear, Forestell leaped from the research vessel into the wake of a female, a calf, and two escort males, hoping to get some good photographs of the animals.

"I *thought* they had dived," he recalls. "In fact, when I jumped in the water, there were the mother and calf ten feet away, her pectoral fin nearly touching my swim fins. A close encounter! When I looked down to snap a couple of pictures, I saw one of the cow's escorts railroading like a freight train straight for me. 'Oh, my God, the escort's going to try to put himself between me and the female,' I thought. Unfortunately, there was no way forty tons of whale meat could fit. I blindly pointed my camera down to take a couple of last pictures. That way, if they ever found the camera they'd be able to piece together how I died. But about thirty feet below me, the whale suddenly stopped, turned slightly on its side exposing one wide-open eye, and began blowing bubbles from its blowhole. They swarmed upward, hit my swim fins, and broke into a blinding screen of fine bubbles. When the bubbles cleared, the whales were gone. It was very controlled, calibrated, and deliberate behavior. It illustrates, I think, the importance of using bubbles skillfully."

If, indeed, song requires the skillful manipulation of air, then these ocean symphonies may be the ultimate measure of a male's skill with bubbles. But in truth, it is not even known how they produce song, since they have no vocal chords. They may manipulate air flowing through their larynx or squeeze it through a series of internal passages, though none has yet been proven to serve this purpose. However it is produced, song may convey a multitude of complex information; "Here's how long I can hold my breath, here's how well I can control air, here's how well I can make bubbles. Because I can make bubbles so well, I'm a more successful feeder and protector. I'm more fit."

A final observation that might tie in with the "song-as-fitness-display" theory is the fact that most males within each breeding population sing more or less the same song. Those in the Atlantic sing their own oceanic anthem, while those in the Pacific sing an altogether different one. Is it possible that once several dominant males establish their credentials through song, and in turn prove successful with females, that other males begin imitating them, thereby leading entire whale populations to sing along? Another male might add an irresistible twist to the song, becoming the new star, in effect, and displacing the old male. Others get on his bandwagon. Stiff competition for females may force the males to improvise. As of yet, no leader has been found, though scientists continue to look for evidence to support the theory.

Admittedly, such musings about the purposes of whale song are difficult to substantiate, and at this point they have hardly even reached the status of hypotheses. To further complicate the matter, the elaborate songs of the humpback may not reflect any actual survival ability of the male at all—they may simply be passed through the genes to his and his mate's offspring. Perhaps, like the tail of a peacock, songs are mostly for show, with little substance behind them. The female's fickle preference for a particular trait may set into motion a trend. Each generation will be selected for individuals who possess those qualities. Among birds, feathers might grow longer or more colorful. Among the humpback, the song may grow more elaborate.

On the other hand—and many biologists insist upon this point—whatever performance the male puts on must ultimately reflect a superior ability to survive. In other words, the female does not act upon a whim; somewhere is the "recognition" that the male is prepared to offer a truly superior set of genes, of which superior singing is a reflection, not the essence. One is left with only one sure conclusion: the possibilities are endlessly intriguing. But perhaps it is such mystery that makes humpback song so compelling a phenomenon.

Song is but one aspect of the whale's elegant reproductive rites, built around winter migrations to tropical breeding grounds after summers spent in the krill-rich feeding waters of the polar regions. Part of the Atlantic population returns to the warm waters off the West Indies for winter calving and mating, while part of the Pacific population gathers off Hawaii. A century ago, the scenic port of Lahaina, on the island of Maui, was an important whaling station. Although humpbacks live in all the oceans, their scientific name *Megaptera novaeangliae,* translated roughly from Latin, means "big-winged New Englander."

Although the reports of sailors who claim to have witnessed whales mating appear in the literature only slightly less frequently—and with about as much reliability—as do sightings of unidentified flying objects, there are no scientifically documented cases of humpbacks mating. In 1951 two Japanese scientists *claimed* to have seen two humpbacks dive and then swim toward each other at great speed, break surface as they met, and belly to belly, lift their entire forward portions from the water, then splash back into the sea with a resounding crash. This was reported to have continued for hours. Scientists put little stock in this description of a mid-air meeting of whales. Nonetheless, similar behavior was reported in 1955 off the east coast of South America by

P.G.V. Altveer, Master of the *Eemland*. Several years before, the stoker of the S.S. *Molenkerk* reported seeing some two hundred sperm whales engaging in similar behavior. But such alleged "sightings" have long been part of a rich mythology surrounding whales, and they should not be confused with actual knowledge, which is meager at best.

And it is impossible to deduce the dynamics of humpback mating from that of other whales, for the species are vastly different. Observers have reported seeing pairs of California gray whales lying horizontally on the water surface, mating belly to belly. Photographer and scuba diver William Curtsinger has swum among courting southern right whales, and describes a male and female stroking each other's side with their fins, until "shifting his great form, belly up, the male's penis emerges, grey-pink in the half light, sways cobra-like, trance-like, and enters her. Belly to belly they come together, moving gracefully, fluidly, forward." But the true nature of mating among humpbacks will remain a mystery until it is actually seen.

Whatever scientists lack in the way of verification, the poet D. H. Lawrence described whale courtship thus:

. . . tremble they with love
and roll with massive, strong desire, like gods.
Then the great bull lies up against his bride
in the blue deep bed of the sea,
as mountain pressing on mountain, in the zest of life:
and out of the inward roaring of the inner red ocean of
whale blood
the long tip reaches strong, intense, like the
maelstrom-tip, and
comes to rest
in the clasp and the soft, wild clutch of a she-whale's
fathomless body.

87

All that has been learned of the whale's keen intelligence and biological sensitivity, perhaps best shown through its song, brings a certain poignancy to the centuries of exploitation they have suffered. In the days of sail, whalers could sometimes drop to their hands and knees, put their ears to the wooden deck, and hear the shrill calls of a humpback reverberating through the hull. When the Paynes first set out to record whale songs in the 1960s, one of their problems was how to follow the sound without noisy interference from the boat, so they chartered a sailboat. The search had come full circle. But for once it was the mysterious song, not the whale, being sought.

Whether insect, bird, or whale, and for whatever purpose song serves, sounds between animals began very long ago. Charles Darwin speculated that even human communication, in its earliest form, served not primarily to facilitate abstract reasoning, but to aid courtship—a way of charming sexual partners as well as impressing any rivals. "Some early progenitor of man probably first used his voice in producing true musical cadences, that is in singing . . . [T]his power would have been especially exerted during the courtship of the sexes," he wrote.

How long ago vocalizations began, no one can confidently say. Michael Bright, in *Animal Language*, proposes that animal sound may have begun with trilobites, creatures that lived on the ocean floor some 500 million years ago. He theorizes that not only were their external skeletons well equipped for making sound, but that they had been around long enough for nature to have given them the equipment to produce it. "But what if, like their distant extant relatives the horseshoe crabs, they mated at night?" he asks. "Could it be that they produced specific signal noises to help find each other in the darkness? We shall probably never know. . . ."

88

7.

Silent Communication

What are we going to do if it turns out that we have pheromones? What on earth would we be doing with such things? With the richness of speech, and all our new devices for communication, why would we want to release odors into the air to convey information about something?

—Lewis Thomas,
The Lives of a Cell

MEDFORD, MASSACHUSETTS, a town of sixty thousand people on the outskirts of Boston, its rows of clapboard houses and suburban monotony broken by a few church spires rising above the treetops, is not the sort of place you would suspect of having given rise to everlasting pestilence. But it was there, in 1869, at 27 Myrtle Street, that the gypsy moth first took up residence in America. Gypsy moths had been brought from southern France by American naturalist Leopold Trouvelot, who hoped to breed them with silkworm moths in order to create a native silk industry after the Civil War had put an end to Confederate cotton as a source of fiber. He reared the caterpillars in a screened enclosure built around shrubs in his garden. Not only did the two species fail to mate, but the gypsy moth larvae managed to escape when a sudden windstorm blew down their enclosure.

As Trouvelot already knew, and as most New Englanders would soon find out, gypsy moths are very good at eating trees. Within a decade after their escape, they had infested about 400 square miles around Medford. A local resident described their depredations twenty years after their escape: "The walks, trees and fences in my yard and the sides of the house were covered with caterpillars. I used to sweep them off with a broom and burn them with kerosene, and in half an hour they were just as bad as ever. There were literally pecks of them. There was not a leaf on my trees. . . . The stench in this place was very bad."

By 1904 over 4,000 square miles were infested, including parts of Maine, New Hampshire, Rhode Island, and Massachusetts. In 1953 over a million and a half acres of forest in New England, New Jersey, and eastern New York and Pennsylvania had been eaten. In 1971 two million acres of forest in the Northeast were destroyed. Ten years later over thirteen million acres had fallen. Today, the moths still conduct their annual tree-stripping rampages. If not brought under control, they may destroy over 100 million acres in the Northeast alone. Another 100 million acres of Appalachian hardwood forests could be destroyed as the moth moves to the south and west.

It goes without saying that gypsy moths are very good at reproducing. One reason they are such prolific breeders is that they have an excellent method of overcoming distance in order to mate. The moth uses what is known as a pheromone, a chemical that bears a specific message for a member of the same species. Scientists have named the gypsy moth's pheromone "disparlure," after the insect's scientific name, *Lymantria dispar*.

The female gypsy moth releases disparlure from a specialized

90

gland on her abdomen, then fans it on its way through the air by flapping her wings. Since the females don't fly, they send out these chemical messages and wait in a tree. They are rarely disappointed. The male's elaborate antennae, each covered by some 15,000 pheromone-sensitive hairs, will fire in the presence of a single molecule of the attractant, and can detect a female from great distances—perhaps up to seven miles away. One 250-trillionth of an ounce of the aphrodisiac will cause neurons in its antenna to fire at a rate of 200 pulses a second, sending its brain the message for the individual to fly upwind. Four millionths of an ounce of disparlure could in theory attract more than a billion males. Unable to remain trained upon the frail molecular pathway from beginning to end, the male may navigate by flying a zigzag course, intersecting the odor plume at intervals, gradually threading its way to the source.

As the male approaches the tree on which the female is located, he may fly rapidly up and down along the length of the trunk, sometimes brushing the bark with his fluttering wings. Once the male spots a female, he accelerates and tilts his body from a forty-five degree searching position until it is nearly horizontal with the ground. If he happens to be in competition with another male, they rapidly thwack each other with their wings until one finally makes it to the female. He then lands and rapidly fans his wings. In some closely related species, the male christens the female with an aphrodisiac from glands under his wings or on his abdomen. The gypsy moths then copulate, often on a leaf, for about forty-five minutes, at which point the female shakes off the male and walks away. Its purpose well served, her scent gland is retracted, while she releases a chemical to neutralize any trace of the pheromone. Within twenty minutes the single, brief moment of her life during which she is attractive is over, and she may crawl a short distance and lay a cluster of anywhere from 150 to 1,000 eggs, unmolested by the hordes of other suitors still searching the airways for a trace of the chemical. The eggs, meanwhile, fall into a state of virtual suspended animation through winter. In early spring, after hatching, the small larvae, weighing less than a milligram each and covered with long, hollow hairs, suspend themselves on long silky threads from the trees, in wait of wind. A breeze of only 1.3 miles an hour will drag them horizontally the whole length of their tethers. Stronger winds snap their anchor lines and carry them aloft. These larvae have been found as far as thirty-five miles away from the nearest site of infestation and have been dragnetted by airplanes at over 2,000 feet, although they generally land near their launching site.

When French naturalist Jean Henri Fabre conducted some of the first experiments with moth attractants in the nineteenth century, he could hardly believe their power. When a female emperor moth was placed in a gauze bag on a table in his study, she had forty male visitors at the window by evening. For the next week, he wrote, "they seemed to take possession of the house." He even hid the moth in a drawer; still males hovered about the spot. In the early 1900s psychiatrist and entomologist Auguste Forel had some pupae of European silk moths in his studio in Lausanne, Switzerland. When the females emerged, they attracted so many males to the windows that urchins gathered in the street below to marvel at the convention of moths.

Not until 1960, when three United States Department of Agriculture scientists squeezed less than a thousandth of an ounce from the abdomens of 200,000 females, was a weak form of the gypsy moth sex attractant isolated. In lab tests, it took less than a trillionth of a microgram to stimulate a male. In a field test, however, when the chemical was distributed by aircraft over Rattlesnake Island in Lake Winnipesaukee, New Hampshire, it proved quite weak and drew few males. Not until 1969 was a pure, potent form of the sex attractant finally isolated from 78,000 abdomen tips at the Pesticide Chemical Research Branch of the USDA, Entomology Research Division. The first batch totaled thirty grams—enough to bait traps, researchers reckoned, for the next 50,000 years. The USDA currently places some 300,000 traps baited with the pheromone disparlure throughout the country to monitor new infestations.

Not surprisingly, the gypsy moth became a prime candidate for annihilation by its own potent chemicals. The general idea was to throw up enough synthetic sex attractant into the air to confuse them—sending males after so many chemical chimeras that their chances of finding real females were lessened. In one such attempt in the 1960s, known as Operation Confusion, granules containing an attractant were dropped over New Hampshire forests. Unfortunately, where the density of the moths was high, such tactics proved ineffective, because males still managed to locate females. Several years later, in 1973, a large-scale pheromonal air assault was conducted against the moths in Massachusetts. Aircraft bombarded twenty-four square miles with microcapsules of the sex attractant. Scientists then counted the number of males who had been lured to ground traps baited with actual females. There were 1,136 males captured in female-baited traps in an untreated area; but only one male managed to locate a female-baited trap in the treated area. The conjugal disruption persisted for some five weeks after the treatment.

Although it wasn't until 1959 that the first pheromone, produced by the commercial silk moth, *Bombyx mori*, was identified, pheromones are now believed to be one of the most ancient and widely used means of communication in the animal kingdom, being employed by some three-fourths of all species. That same year the term "pheromone" was coined by P. Karlson and M. Luscher, who joined the Greek roots *pherein* (to transfer) to *hormon* (to excite).

Perhaps the only kind of animal among which pheromones are unknown are birds. Pheromone users include the single-cell slime molds and yeast, as well as hamsters, dogs, horses, mice, moose, deer, marmosets and other primates, porcupines, goats, bees, and butterflies. Rabbits secrete them from anal, chin, and groin glands; the female red-sided garter snake through her skin; deer from glands above the eyes.

In many species, pheromones are carried in the urine. When a dog sniffs a lamppost and then urinates, he is plugging into a kind of canine chemical information center, picking up a message from another dog, then leaving his own: FIDO WAS HERE. In lorises, urine-borne pheromones act as scent trails between sleeping and feeding sites. The male "washes" by urinating into one hand, then rubbing it onto the other and onto the feet. He then tracks his urine through the forest canopy, creating an arboreal scentway.

In some species, pheromones are carried in the excrement. Black rhinos, for example, build piles of dung in their home territory. When another rhino approaches, the visitor sniffs, then often sweeps the pile with its horn and stomps through it. Finally, it defecates, kicking and scattering the feces with its hind legs and tracking it with him when he departs. Later, another rhino reaching this malodorous intersection may pick up the trail and follow.

Not all pheromones are related to courtship, and fewer still serve as "sex attractants." So-called aggregation pheromones draw animals together. Slime molds, for example, use them to call to order the microscopic meetings of single-cell amoebae that aggregate into the colonies of slime. When one begins to emit a pheromone pulse, others rally around it, eventually amassing into a multicellular organism. It is one group of related organisms that is responsible for the strikingly beautiful white, yellow, or red etchings often found on the underside of decaying logs.

Still other pheromones cause animals to disperse. Some ant species employ a pheromone that has been called a "propaganda substance," because they use it to wreak havoc among the colonies they invade in search of slaves to take care of menial chores such as raising young and foraging for food. During these slave-taking raids, the pheromones emitted by the invaders cause the victims to panic, while simultaneously signaling the home colony to send reinforcements.

While the specific chemical responsible wasn't identified until the 1950s, in 1609 Charles Butler showed a keen appreciation for whatever it is that caused other bees to swarm when one bee stings a person. He wrote in *The Feminine Monarchie:*

> When you are stung, or any in the company, yea though a Bee have strike but your clothes, specially in hot weather, you were best be packing as fast as you can: for the other Bees smelling the ranke favour of the poison cast out with the sting will come about you as thicke as haile: so that fitly and lively did he express the multitudes & fierceness of his enemies that said They came about me like Bees.

Still other pheromones act as passwords, identifying either an individual's sex or rank in a group. The tiny desert wood louse defends its burrow from all but its mate, which announces its arrival with a "welcome" pheromone. Other pheromones incite aggression. The odor of a strange male guinea pig or rabbit, for example, provokes attacks by other males. At times the chemical itself becomes the agent of attack. In "stink fights" between ringtailed lemur males, combatants rub a pheromone secreted by their wrist glands onto their tails, which they then wave menacingly at each other like wands of destruction.

But it is those pheromones that play a role in courtship, in particular the sex attractants, that have captured much public and scientific attention.

In 1971 scientists in England discovered that if they took a certain chemical produced in the testes of a boar and sprayed it near a sow in heat, she would spread her legs, arch her back, and perk up her ears in a rigid mating stance. This so-called "immobilization reflex" is usually reserved for a courting boar in the privacy of the sty or barnyard, and the chemical that prompts it is a pheromone called "alpha androstenone." In nature the musky-smelling pheromone is transferred from a boar's testes to his saliva, then to the sow during courtship, during which the male foams at the mouth.

Pig courtship in itself is relevant enough to merit some description. In addition to salivating heavily, the courting male pig grinds his teeth by moving his jaws side to side as he pokes his snout between the female's hind legs, sometimes lifting her hindquarters off the ground. Frequently the boar sings a mating song, or *chant de cour,* a series of rhythmic, baritone, eighty-five- to ninety-five-decibel grunts. Emitting them as fast as a half dozen a second, his lungs are so taxed that he must pause every fifteen or twenty seconds to catch his breath. The female, meanwhile, nuzzles the male's scrotum, flanks, and genitals, bites his ears, and sometimes tries to mount him. After the pheromone has taken effect, she goes into the mating stance, becoming so rigid that she hardly budges as she is mounted by the boar. A synthetic form of the pheromone is now sold in an aerosol dispenser under the name "Boar Mate" and is used to render sows more amenable to artificial insemination.

The discovery of the pheromone alpha androstenone provided an insight into the moving spirit of pig courtship that might have remained down on the farm had a similar chemical not also been found, in 1974, to exist in minute quantities in the armpits of humans. What's more, the chemical existed in both species in the same ratio to the male sex hormone testosterone, suggesting that the pheromone might play a similar role in humans. While most people can appreciate the chemical kinship of all life, there is something wildly amusing in the possibility that nature may have used some of the same pheromonal nuts and bolts in assembling the sexual behavior of man and his most berated domestic animal.

In 1981 still another chemical link between the interests of men and pigs was discovered: truffles contain large amounts of alpha androstenone, which helps to explain why pigs find the fungus so delectable and why they have been used for centuries in Europe to help truffle hunters locate them in the woods. Pigs, in fact, have been known to root out the fungi from a meter deep in the ground. "The biological role of this boar sex pheromone might explain the efficient interest of pigs in search of this delicacy," concluded the scientists who made the discovery. Perhaps coincidentally, truffles have long been thought to have aphrodisiacal qualities. It is said that Napoleon Bonaparte, concerned that he had not provided France with any male heirs, began eating them in quantities. The future king of Rome was born shortly afterward—or so the legend goes.

It was once thought that pheromones, especially those that helped attract one sex to the other, would never be found outside the insect world. Once pheromones were discovered in reptiles, scientists expressed skepticism that they would ever be found in mammals. Once verified in mice and some other rodents, researchers were confident the trail would end there.

In 1971 five different substances—fatty acids known as "copulins"—isolated from the vaginal area of female rhesus monkeys were claimed by the investigating scientists to increase sexual activity in males. A synthetic concoction of these acids was then dabbed on the genital area of unreceptive female monkeys. Males responded as if the females were actually in heat. In one experiment, a trained rhesus once pressed a lever 250 times to win one such anointed female. When gauze was stuffed in their noses, however, the males were unable to distinguish between receptive and unreceptive females. The experiments suggested that monkeys actually possessed sex attractants. Not long after these findings, the same fatty acids were discovered in women. But by then the primate studies had begun to wither under stinging criticism. For one thing, the copulins were never compared against a range of other substances that might have had similar effects, thereby disqualifying claims to any special properties of the copulins. For all anyone knew, the fatty acids may have been a by-product of a physiological change that prompted solicitiousness in females. The copulins could actually have been odorous cues to males that the females were ready to mate. In short, effects were never clearly linked to causes. In retrospect, the main significance of these controversial copulin experiments was to suggest a role for odor in primate courtship and to generate interest—as did the swine experiment—in research into the possible role of pheromones in human courtship. It has now been demonstrated, for example, that air-borne chemicals can have a major regulatory effect upon the menstrual cycle in women—a claim that, twenty years ago, the scientific community would have dismissed as pure alchemy.

While for the most part scientists remain skeptical about the role pheromones play in human behavior, people continue to employ various substances, as they have for thousands of years, in the hope of making themselves attractive to the opposite sex. Musk, traditionally obtained from a walnut-sized gland found in the navel of the Himalayan male musk deer, has been used in perfumes for more than 3,000 years. Until recently, men in Melanesian Islands of the Pacific wore a supposedly seductive, musky-smelling native leaf when they danced, while women rubbed into their hair an aphrodisiac of coconut oil mixed with turmeric. In the classic Chinese drama *The Transmigration of Yo-chow,* a young doctor addresses a poem to his fiancée:

When I climb to the bushy summit of Mount Chao,
I have still not reached to the level of your odorous armpit
I must needs mount to the sky
Before the breeze brings to me
The perfume of that embalsamed nest.

An early scientific proponent of the ubiquitous role of pheromones in humans was physician Alex Comfort, who wrote in the
prestigious journal *Nature,* "Humans have a complete set of organs
which are traditionally described as non-functional, but which, if
seen in any other mammal, would be recognized as part of a pheromone system." These include the apocrine glands, some of which
do not produce sweat and so probably secrete something else.
Such glands are usually located near conspicuous hair tufts which
may serve to disperse the chemical into the air. Humans have such
glands in the armpits and pubic areas. Vestigial scent glands may
also exist on the feet, leading some to speculate that there might be
more to the proverbial foot fetish than a simple obsession. Comfort
also describes a "characteristic, powerful and pleasant axillary odor
emanated only by women—some of them at all times, so that their
presence in a room can be recognized, and others only occasionally. This odor appears to be unrelated to sexual excitement, but is
itself an attractant. It does not appear to be produced by males."

Comfort's belief was lent some credence by a series of experiments that attempted to document the influence of the swine pheromone, alpha androstenol, on human beings. While most of the
tests proved highly suggestive, none produced the air of tight con-
clusiveness that would earn a scientific stamp of approval. In one
of the experiments, a group of first-year psychology students was
given brief written profiles of six applicants for a position on the
Students' Representative Council of a large university and asked to
rate the candidates in a number of specific areas. The subjects

were also given surgical masks impregnated either with androstenol, a second compound, or nothing at all. Results showed that females under the influence of alpha androstenol consistently rated males more favorably in certain areas, while giving unfavorable ratings to females. "We have apparently demonstrated, possibly for the first time in a controlled experimental situation, that a complex human behavior can be influenced by the odor of substances known to have phenomenal properties in animals," the authors concluded. In a later experiment, conducted by David Benton, the compound was rubbed on women's lips in the middle of their monthly cycle; women so treated rated their moods as submissive rather than aggressive more frequently than did those tested with a placebo. The compound's effect seemed to be greatest at peak fertility—a time when women are also known to become highly sensitive to musky-smelling compounds.

Not surprisingly, some of the strongest scientific advocates for the presence of human sex attractants are found in the perfume industry, which spends millions of dollars annually in search of natural attractants. With a flourishing multimillion-dollar market for aphrodisiacs that don't work—everything from ground rhino horn to rare herbs—anything that could be marketed as a scientifically verified sex pheromone would create a tumult at the cosmetic counters of Bloomingdale's and Saks Fifth Avenue. In 1981 Jovan introduced their androstenol-spiked Andron perfumes and colognes. With consumers perhaps fantasizing that the human equivalent of "Boar Mate" had been discovered, sales in the first year reached $18 million, making it the best-selling Jovan perfume ever. Whether the pheromonal additive works because pheromones really do affect people, whether it works because people think it works, or whether it doesn't work, is not entirely the point. People have anointed their bodies with potions derived from animals for thousands of years. Many of them have a musky aroma reminiscent of alpha androstenol. The point is, according to people in the industry, that the aromas carry certain "rewards" for the user, making one *feel* sexy if not actually increasing one's sexiness.

Like many fragrances used by perfume companies, the potion containing alpha androstenol was manufactured by a subcontractor—in this case International Flavors and Fragrances, Inc., of New York City. IFF also supplies the aroma in Downy Fabric Softener, the scent in Colgate's Irish Spring soap, and a couple of more exotic smell sensations, such as essence of slum (a blend of garbage and urine) for a Smithsonian exhibit, odor of cave for an amusement park, and the fresh leather smell of new cars.

Dr. Craig Warren, vice president and director of organoleptic research at IFF, was one of the researchers who led the search for the synthetic compound used in Jovan's Andron line, research that actually began in the late 1960s. While downplaying any Casanovian claims, Dr. Warren is careful not to disavow Andron's potential as an attractant. "What we supplied to Jovan was the first perfume with androstenol, although it has musk compounds as well. For lack of a complete scientific picture, entire anecdotal histories have gone into this perfume. But the combination of the androstenol, musk, civet, and other materials in the fragrance is probably stimulating the hypothalamus to produce the right kind of hormones which cause a person to feel sexy—testosterone in males, perhaps progesterones in females. It's not magic, as many people seem to think. It's biochemistry."

According to Dr. Warren, one of the most striking features of scents—and musk in particular—is that out of thousands of aromas that people find pleasant, musk has been sought after as a precious material for centuries. Part of its appeal may be learned, he concedes. But why, out of the thousands of odorous materials available to people, would so many cultures independently zero in on musk? Perhaps it really does provide some "hidden reward," says Warren.

There is now circumstantial evidence from studies conducted at Duke University that musk functions as a sexually related pheromone in humans. Using a panel of women, researchers noticed that the length of the women's menstrual cycles was shortened from about twenty-eight or twenty-nine days to about twenty-four or twenty-five days by exposure to musky odorants. This was the first anecdotal evidence suggesting that, in addition to its pleasing, sexy smell and hedonic tone, musk may actually affect sexual behavior in some very subtle ways. "As a way of communicating our preference for a particular odor, we say it smells good, or smells sexy," says Warren. "All those words are very crude approximations of what may be actually occurring in our brains—because the odorant might be causing changes in that part of the brain that controls the production of the sex hormones. This is the hidden reward."

The phenomenon of smell is rooted deeply within an ancient part of the human brain. While vision and hearing are handled by the more recently developed cortex, smell is situated deep in the center of the brain in a thumb-sized organ nestled atop the brain stem—the hypothalamus. This section mediates many basic bodily functions such as body temperature, blood sugar, and blood pressure. It also mediates the pituitary, which controls hormones. In short, the olfactory mechanism is perfectly sited to allow information to go directly from the nose to the part of the brain that controls sexual behavior. "The striking thing is that we are biologically set up to convert olfactory signals into changes in emotion or sexual behavior," says Warren. "This change certainly fits within the context of the definition of a pheromone. Even though one effect is small, we have been using fragrances for the last five thousand years to attract people to us."

While musk has one of the most venerable histories, the molecules that have gotten the most press recently are the urinous-smelling steroids such as the androstenones. In perfumes, the ac-

102

tive ingredient androstenone is often converted to an alcohol-based form that smells sandlewoody, slightly musky, but, in the words of professional smellers, still has some "urine notes" to it. Its proven effect on pigs gained it a large scientific following, but no work has yet been done that proves, unequivocally, that it serves as a pheromone among humans. Still, says Warren, "it is a very interesting molecule." Certainly, the research results that have been obtained with this odorant are highly suggestive.

The term "sex pheromone" can mean very different things. A sex pheromone may act as a sex attractant or as a regulator of the reproductive cycle, one having nothing to do with the other. While human sex pheromones have yet to be proven attractants, they have been shown to influence reproductive functions. Some of the first solid evidence in this regard was the discovery that women living together in college dormitories tend to synchronize periods. In 1970 Martha K. McClintock of Harvard University asked 135 women in a suburban women's college to note the onset of their periods as well as the time spent with males throughout the semester. It was found that roommates and close friends had the highest frequency of synchronized cycles. Furthermore, women who reported seeing men less than three times a week had significantly longer cycles than those who saw men three or more times. One woman reported having a cycle of six months, but when she began spending time regularly with men, it shortened to four and a half weeks. When she then stopped seeing men, it again lengthened.

While pheromones had now been indisputably implicated, the source of the chemicals responsible for the "McClintock Effect" has only recently been found. In 1985 Dr. George Preti of the Monell Chemical Senses Center and Dr. Winnifred Cutler of the University of Pennsylvania achieved the same effects by using secretions from the underarms of men and women. Preti and his colleagues asked men and women who served as secretion donors to stop using deodorants and deodorant soap for several weeks, then to wear special absorbent cotton pads under their arms for six to eight hours at a stretch. At the end of the day, the subjects placed the pads in jars and froze them to prevent bacterial growth. Preti had the jars collected regularly and the secretions in the pads extracted. The extracts were then rubbed under the noses of another group of female subjects. Over the course of several months, females who had a history of irregular cycle lengths, tended, when subjected to the extracts of the males, to regularize their cycles at about 29.5 days, give or take a few days. The 29.5-day length is significant because it is known that a cycle of this duration is correlated with the highest fertility in women. Furthermore, when subjected to extracts of women, women with regular-length cycles tended to synchronize their cycles with those of the donors.

"We feel that the substance responsible is produced by interaction of secretions from the apocrine glands in the underarm with bacteria," says Preti. "But we aren't sure yet if the active ingredient is actually androstenol. The hypothesis is that there's some organic compound produced in the underarm that affects the endocrine system by way of smell. Our next step is to isolate various components of the substance we took from the armpits. Our experiments so far offer few concrete answers, but they do raise thousands of new questions."

If one of the substances under study—say, alpha androstenol—indeed proves to influence the reproductive cycle in women, then one could claim that a pheromone has at last been discovered within that group of animals where many least expected it to occur. And as has long been known to be the case in other animals, human beings will be faced with the fascinating conclusion that their own reproductive cycles can be influenced by odors and by the mere presence of a partner.

8.

Unconventional
Courtship

Nature grants no monopolies in resourcefulness. She does not
even seem to hold much with the notion of portioning it out
hierarchically. Gold, she decrees, is where you find it.

—David Quammen,
Natural Acts

DURING HER RESEARCH at Gombe Stream in Tanzania, pri-
matologist Jane Goodall once saw an enterprising chimp named
Mike pick up an empty kerosene can near the camp, hold it in one
hand, and drag it noisily along the ground as he stormed through
his troop. This "charging display" helps determine a chimp's rank
in the troop. But Mike was onto something new. A weakling, he
had always ranked near the bottom and stood little chance against
the dominant male Goliath—until, that is, he employed some
technological know-how. Mike soon learned to use three empty
cans in the performance, jingling them in front of him as he
charged. Even the mighty Goliath quickly paid Mike homage;

panting nervously, he bowed to, then kissed and groomed Mike. Mike soon ascended to the dominant position in the troop, entitling him to more females, more food, and many other perks—all of which he won, not by strength, but by virtue of cleverness.

Like a new river across a plain, nature may cut many routes to the same destination—in this case to dominance. Mike's invention suggests that the "fittest" does not necessarily mean the strongest physically. Cleverness and ingenuity must be factored into the equation that determines which males gain the females. What's more, the idea that there are often several routes to the same end grants animals a greater degree of individuality in solving the challenges of reproducing than has been traditionally believed. There are both conventional and unconventional means.

Traditional biology took a rather one-dimensional view of male dominance: only the strongest males gained access to females, while the rest remained unmated. In reality, even among the most polygamous societies, where a few males hoard almost all the females, subordinate males may, through deceit or trickery, have at least some chance of fertilizing a female. Ranging from the sublime to the violent, these behavioral options are often referred to collectively as "alternative strategies."

One of the most common is known as the "satellite strategy," in which males mill around the borders of the dominant male's territory and attempt to intercept females passing to and from it. Most male North American green tree frogs find mates by calling repeatedly from a pond. But about one in seven males stations himself about a half meter from a calling male and intercepts the females as they arrive. It's a highly successful technique, though the frog that does all the work sometimes butts or fights the interloper or tries to frighten him away with an aggressive call. An added benefit of this strategy is that silent satellites do not expose themselves to the dangers of bats or other predators. While the dominant male does the dangerous advertising, the satellite reaps the rewards.

Even more interesting, some animals may be satellites one day and territory-holders the next. The white-faced dragonfly, for example, seems to have two equally workable alternatives for finding a female. It can either claim a small patch of pond that offers a perch (perhaps algae or some underwater plants where eggs can be laid), or it can patrol the pond's edge and ambush females. To study the relative mating success of these two classes of males, Syracuse University biologists Larry Wolf and Ed Waltz observed dragonflies at a small pond near Jamesville, New York. Painting the wings of males and tethering females to nest sites with light-weight monofilament fishing line, they recorded the relative mating success of males employing each tactic, and then determined which one resulted in the highest number of fertilized eggs. They found that transient male dragonflies, freed from the burden of protecting their turf, had more time to search a wider area for females and, once they found one, more time to mate. They could consequently fertilize a higher proportion of eggs. But this advantage was apparently offset by their lack of a safe territory in which the female could lay her eggs. Thus, the average mating success of territorial and transient males is more nearly equal than many other systems of alternative tactics.

This finding was somewhat surprising because traditional biology would have assumed that territory-holders were fitter than transients—perhaps bigger or better in aerial combat—as a result of which they could lay claims to better areas, and hence produce more young. Even more intriguing was the discovery that the same males employed different strategies at different times. The factors that lead a particular male to choose transience one day and territoriality the next are not known, but the mere existence of options makes it clear that even the primitive dragonfly lives in a dynamic interplay within an environment of tremendous flux and complexity.

Many animals use more active means of gaining leverage over a successful male. Sexual interference, in which one male actively disrupts the courtship or mating of another, is based on the apparent premise that one can increase one's own chances of mating by decreasing another's. In 1981 zoologist David L. Martin documented a possible example of interference among snakes while studying cottonmouths at Lake Martin, in Louisiana's St. Martin Parish. Lake Martin is a swampy area with a central patch of open water surrounded by thick cypress and tupelo trees. Faced with the difficulties of monitoring his subjects in difficult terrain, Martin rigged a few snakes with radio transmitters, so when one was on the move, he could locate it by homing in on the strength of its signal with a radio receiver. At 7:57 one summer evening he was tracking a female as she slithered slowly along some floating vegetation. Soon a male picked up her trail and began following her. Within a few minutes the male had reached the female and they began a courtship ritual, which included tongue flicking, convulsive movements of the body, and sliding his chin along her spine.

At 8:59 a third snake—a male interloper—entered the scene. Perhaps he had picked up the female's scent, aware that she was already taken. In any case, neither the female nor her suitor saw the interloper coming, and not until he actually bumped the female's tail with his chin did she react by jumping away and alerting her partner. He turned suddenly and confronted the second male. Facing off about two inches apart, they humped the front thirds of their bodies in unison, in apparent preparation for a combat dance. While each species is unique, this "dance" generally involves pushing each other, entwining, or even flipping the opponent. In the case of the two males, the mere threat of bodily contact drove off the interloper. After pursuing him a short distance, the first male returned to the female, and as night fell, resumed courtship.

Like alligators, snakes generally rely heavily on tactile communication during courtship, with the male performing most of the movements. He first attempts to pacify the female by gently rubbing his head over her back. While the motions often appear chaotic, all are directed toward ultimately bringing their bodies in position to mate. Among some species, the male lies across the female's posterior and strokes, scratches, or vibrates against her cloacal region to induce her to twist it toward him. Many nonvenomous male snakes follow, chase, intertwine, rub chins, or stimulate the female with wavelike movements of the body.

A subtle interference technique is employed by the American tiger salamander, the male of which can dislodge another and take his place, apparently without his knowledge. The female initiates courtship by nudging the base of the male's tail, who then leads her in a slow and deliberate walk, with her head nudging his tail. As they walk, the male deposits on the ground a tiny sperm packet, or spermatophore, then leads the female over it so that it attaches to her vent. But an interloper can slowly slip between the male and female, like an overeager movie-goer sneaking into line outside a theatre. Because the male interloper nudges the leading male's tail just as the female did, the leading male will be unaware that anything is amiss. (The female, meanwhile, has slid along the interloper's side and has begun following behind him.) The leading male deposits his sperm pouch on the ground, and the interloper then drops his on top of it as he passes by. The female picks up the top one, leaving the original deposit behind. The salamanders then part, with the original male unaware of having been deceived.

One sometimes hears about the phenomenon of the powerful corporate executive who surrounds himself with weaker colleagues in order to minimize challenges to his authority. Among some animals, the dominant male will tolerate the presence of the subordinate only if the weaker submits to copulation.

In the 1970s zoologist Robert Trivers spent eight months in Jamaica studying the highly territorial lizard *Anolis garmani*. He captured and marked adults from a nine-acre area outside of St. Elizabeth, lush with pimento and mango trees, released them, and began recording their behavior. He found that dominant males controlled large territories that sometimes included several trees. Such a male maintained exclusive access to the female residing within his territory by aggressively fighting most trespassers. Raising his nuchal crest, throat inflated, body fully extended, a dominant male would lunge repeatedly at an interloper, attempting to bite him on the jaws, neck, and side. Not surprisingly, dominant males accounted for ninety percent of the copulations in their territory.

What about the other ten percent? Trivers found that nondominant males employed any of several alternative strategies. In one case, a smaller male hid in a corner of the territory and managed, probably without the dominant male's knowledge, to slip onto a mango tree and copulate with a resident female. In a second case, the interloper brazenly marched to the pimento tree at the center of the territory and mated. In yet another case, the interloper copulated with the female, then returned later, probably while the dominant male was visiting females in one of the eight small trees adjoining his territory. But perhaps most interesting, Trivers observed that a smaller male actually occupied the same territory as the larger male, living in such subordinance that he occasionally was mounted by the dominant male—an event that literally turned him black with fear. (As a human often turns red with embarrassment, a male lizard turns black when frightened.) Apparently, the dominant male tolerated the presence of the smaller, submissive one and perhaps even benefited from his presence. In turn, the submissive male reaped the benefits of a lush territory and probably even managed to mate occasionally with resident females. As Trivers concluded, "An occasional buggery might be a small price to pay for the advantages of remaining within the large male's territory."

In 1985 two scientists from the University of Texas discovered that some male red-sided garter snakes masquerade as females in order to slip past the defenses of other males and thereby gain a significant edge over the competition. Every spring these red-sided snakes come together in great congregations called "mating balls," in which a single female may be surrounded by a writhing ball of up to a hundred males. Only one will successfully mate with her. But upon close inspection, the researchers discovered that some of the mating balls were comprised entirely of males, one of which was actively solicited by the others. They found that the "she-male" released a powerful phermone that mimicked a female's scent. The counterfeit perfume could prove an invaluable asset to a male, who could use it to lure suitors away from a female and, in effect, clear the way for himself. Experiments showed that in twenty-nine out of forty-two tests, the impostor indeed reached a female first.

In one manner or another, mimicry is used by a wide range of animals. The male stickleback fish will sometimes mimic the behavior of a female, be granted admission to a male's harem, then fertilize the eggs. The male North American hanging fly is also a gifted mimic. During a normal courtship, a male captures an insect and presents it to a female, then mates while she is eating it. Some males have mastered the art of mimicking female behavior, thereby coaxing a male out of his gift. Once they have received the gift, they take it and go in search of a female. This not only puts the other male at a disadvantage, but saves the mimicker from the risky task of capturing a gift, during which time he must expose himself to the ever present danger of spider webs.

In some cases, predators use mimicry, not to attract a mate, but to lure a victim. Females of the carnivorous *Photuris* firefly have garnered a reputation as the *femmes fatales* of the insect world. Males of this species recognize females by a particular sequence of flashes. One species, for example, blinks two or three pulses at 1.2- to 1.4-second intervals and repeats these pulses every ten to fourteen seconds. So specific is the signal that males will ignore pulses of light separated from one another by about two seconds and repeated every four to seven seconds.

University of Florida scientist James Lloyd once "communicated" with various species of fireflies by mimicking their codes with a flashlight. When he spotted a male and imitated the female's response to that species, the male drew near. After some arduous stalking of fireflies in Virginia, Illinois, and Florida, he discovered that the carnivorous *Photuris* females mimicked the female courtship code of a closely related species. It did not take Lloyd long to discover to what purpose these counterfeit signals were put. One evening he set out, flashlight in hand, in search of *Photinus consanguineus*, whose code consists of two short pulses separated from one another by an interval of about two seconds and repeated about every four to seven seconds. "While searching for female, I received a response from the direction of a low weed along a stream," he wrote. But when he investigated he found a large black *Photuris* female. "I watched this female for the next half-hour, and during that time she responded to twelve passing males of the Photinus species." (Of these, one flew into the stream, two flew into the grass below her and dropped out of sight, while the other eight flew to within a meter of her.) "The last male attracted, after three or four flash-exchanges, landed about 7 cm from her. After another flash sequence I turned on my light and found him 15 cm from her. . . . Following the next flash exchange, after a pause of 10 to 15 seconds, I checked and found she was clasping him and chewing on his pronotum."

The satellite strategy, sexual interference, submission, and mimicry are but a few of the unorthodox means that underdogs may use in order to mate. Sometimes it is as simple as sneaking copulation. The male blue-headed wrasse, a colorful fish of the Caribbean, usually defends a territory and judiciously courts females who visit it. Often, as the courtship between a dominant male and his females reaches its climax and gametes are shed into the surrounding sea, a subordinate male rushes to the scene, releases milt, and flees, leaving behind at least a portion of eggs fertilized with his sperm.

Olive baboons have developed an elaborate technique for dethroning the dominant male. In what is known as the consort strategy, an unmated male will strike an alliance with another single male, and together they will attack the dominant male. The one who initiated the attack wins the female. Later, the other participant may initiate yet another alliance, this time dethroning his former ally.

Subordinate males of some species resort to "rape," which constitutes an alternative strategy insofar as it may lead to insemination of a female. Many scientists object to the term "rape" because it is a loaded word, often provoking human sentiments that may not be relevant to the behavior of other animals. Some prefer instead to call it "forced insemination." Others prefer the terms "resisted mating" or even "willing resistance." Whatever the terminology, however, it is clearly an act done against the female's will. A group of male mallards will sometimes attack a paired female, who may kill herself in a panicked attempt to escape. Observers once witnessed six males attacking a blue-winged teal in Manitoba. With the female's escape cut off in every direction, she dove into a lake to escape and was not seen to resurface.

Alternative strategies are more than just a curious sideshow to the mainstream of animal behavior. They raise important questions about the role of the underdog. At one time it was thought that subordinate males were little more than spares to replace killed, injured, or retiring dominant males. Traditional thinking has all but ignored their wider contributions. Far from being leftovers, they may actually be among the evolutionary elite, for it is often these "outcasts" who emigrate to new areas. They become the pioneers, staking out and adapting to new habitats. Most mammals, for example, are reluctant to explore unfamiliar territory. It is the outcasts—those who are forced out of the best habitats, as a result of their lowly status—who disperse and adapt to new ones. In the process they unwittingly carry the genes of the new colonists. Thus, subordinates may provide the impetus for major evolutionary changes among animals. In the words of Edward O. Wilson, "outcasts are at the cutting edge of evolution."

This theory suggests an explanation for the evolution of the fourteen species of finches in the Galapagos Islands. The dominant birds among the original colonists probably staked out the territory most like their native habitat, while subordinate individuals of subsequent generations were forced into successively more unfamiliar habitats. Gradually, each group adapted. The weight of recolonization may have fallen largely upon the subordinates.

Within a species with which we are unfamiliar, individual animals may seem to look and act the same. Often it takes a personal familiarity to be able to distinguish all but the most glaring differences between animals. Yet individual differences abound. The fact that animals may follow different routes to winning a female— even change their strategy from one day to the next—forces one to question not only the rigid boundaries that have traditionally circumscribed animal behavior, but the ultimate value of the underdog as well.

9.

The Monogamous
Way of Birds

BORN IN 1946, George Archibald grew up in the bayside village of
Sherbrooke, Nova Scotia, about twenty-five miles northeast of
Ecum Secum. Situated on a small cove, Sherbrooke marks the
inland end of the five-mile-long estuary of the St. Mary's River.
With water everywhere, the area is rich in birds. Eagles, osprey,
cormorants, and American mergansers fished the river in front of
the Archibald house. As a child, George often accompanied the
local fishermen to outlying islands, where he explored for nests of
eiders, terns, and gulls.

He maintained a keen interest in birds throughout high school
and college, and by his late twenties a major goal of his life was to
work with one of the most endangered families of birds in the
world, the cranes. Seven of the fifteen crane species and sub-spe-
cies are threatened with extinction, including the legendary
whooping crane of the United States and Canada. These, like
many other species, have been hunted mercilessly, and their wetlands have been destroyed. Whoopers, which are still among the
rarest of all cranes, numbered only about seventy-five at the time
Archibald first became interested in them. The most famous

whooper was a female named Tex, who had been hatched and reared at the San Antonio Zoo in 1967. (Tex had been mistakenly identified as a male, hence her masculine name.)

Tex's parents, Crip and Rosie, were descendants of a once prominent flock of cranes that summered on the gulf coast of Texas and wintered in Canada's far north. In 1967, at the San Antonio Zoo, they gave birth to Tex. A sickly hatchling, Tex spent the first six weeks of his life at the home of the director of the San Antonio Zoo. Rather than leaving the young whooper to become little more than a curiosity at the zoo, as her parents had been, zoo officials decided that she should be one of the pioneers in a captive breeding program, with her own offspring being released into the wild. As a result, when she was several months old, Tex was shipped to Patuxent Wildlife Research Center in Maryland, home of the U.S. Fish and Wildlife Service's ambitious endangered species program, which biologists were just beginning to set up. Her intended mate would be Can-Us, an injured whooper who had been caught in Canada and brought to the United States.

Unfortunately, Tex had a problem. Since she had been raised by people, she had never learned to associate socially or sexually with other cranes. Lacking the steadying and stimulating companionship of a male to prompt the secretion of hormones that would trigger ovulation, Tex was also barren. Tex was a victim of a phenomenon known as imprinting, in which new-born animals tend to identify with whatever object is nearest to them during the first few days of life. If a duck is born in the company of a dog, it will seek a dog's company. A crane raised by humans will seek humans, both for companionship and as sexual objects.

In 1969, about two years after Tex was sent to Maryland, Archibald, then a graduate student in ornithology at Cornell University, arrived at Patuxent to study whooping cranes. After some preliminary observation of Tex, he quickly realized that she preferred humans to Can-Us. Archibald soon got the idea of becoming her mate, dancing with her and providing companionship until she ovulated—at which time the relatively new technology of artificial insemination could be used to fertilize her eggs.

Archibald and a colleague, Ron Sauey, meanwhile, had begun devising plans for a special crane-breeding facility in Wisconsin, where they could bring together many of the world's endangered species of cranes. Set up on an old horse-breeding farm that belonged to Sauey's parents, in Baraboo, about an hour north of Madison, it was called the International Crane Foundation. Archibald's idea was to bring Tex to the crane-breeding facility, where he would attempt to form a human pair-bond with her by mimicking a wild male crane's courtship dance and thereby entice her to release a few ova, which could then be artificially inseminated with semen taken from a male at Patuxent.

People at Patuxent were skeptical. Securing sperm from a male crane was a sensitive procedure that had taken years to perfect. Once a donor male had become accustomed to humans, he would respond to their stroking his thighs—something that accomplished the same thing as foreplay among two cranes. Then a few drops of sperm could be collected in a small glass vial pressed against his cloaca. The sperm would then be flown to Wisconsin. There was consideration at the time of preserving the semen by mixing it with a special suspending agent, then freezing it in liquid nitrogen at 196 degrees below zero centigrade. Later, when a female crane was available, it could be thawed and used to inseminate the female. Theoretically, a male who had been dead for years could thus be genetically resurrected to father offspring. The technique was developed by Dr. George Gee, a Patuxent scientist whose name figures in the final part of the story.

In 1976 Tex was crated, put on a commercial airliner, and flown to Madison. Archibald met her at the airport and took her to Baraboo, where he released her into a small shelter connected to a large outdoor paddock. The enclosure was divided by a screen;

one side was Archibald's, the other belonged to Tex. George moved his bed and desk inside the paddock for the length of the six-week breeding season. For the first time in her life, Tex began to act as if she had not only a steady companion but her own territory. George would become a permanent fixture for a crane who had previously known only transience and insecurity.

George and Tex danced a great deal. Wild cranes dance a lot early in a relationship when insecurities run high. Facing Archibald, Tex would flap her wings and bob her head up and down. George would then rise up on his toes and bob up and down in a syncopated rhythm: when she bowed, he rose, and vice versa. He would also flap his arms. If ever he slipped out of rhythm, Tex would threaten to abandon the dance. In between dances, the two of them would charge across the paddock side by side. Then they would dance some more until Tex spread her wings, turned around, and raised up on her black toes in a solicitation posture. At this point a wild male crane would have flapped its great wings and jumped up on her back, maneuvering with swift wing beats until their cloacas met and semen was deposited. Instead, George summoned the AI team. Two assistants would approach the site, syringe in hand, and fight past Tex's stabbing beak and flailing feet, then hide behind the enclosure. Tex would rush over to George and begin a victorious unison call in celebration at having driven off the intruders. Archibald would engage the excited Tex in more dance, and when she went into her solicitation posture, the assistants rushed out with a semen-filled syringe and quickly inseminated her.

In 1977 Tex laid her first egg. It was infertile. The following year she laid her first fertile egg. Four weeks later, Archibald heard a peeping through the thin shell, but for unknown reasons, the chick died a short time later. The next few years were trying ones. Tex laid another egg that was so soft-shelled it broke. In 1980 and 1981 Archibald was so preoccupied with other matters that he spent little time with Tex, and she refused to lay an egg at all. By 1982 he was discouraged and decided to court her once more. If that failed, he would consider the project at an end. That final spring he released Tex from the paddock, hoping that freedom would inspire her. Tex immediately took up residence in the middle of a pasture and built a nest—a casual mound of twigs and leaves. Archibald moved a small wooden shack to the middle of the field. He kept Tex in the shack at night and arrived at the field at dawn every day to release the crane from it. Archibald whiled away the days reading and resting inside the shack. Tex often stuck her head up to the frost-glazed window and purred, signaling Archibald to come outside.

George learned the meaning of all of Tex's calls and movements. A low rumbling sound meant she wanted to go for a walk. Hands dug into the pockets of his favorite corduroy overalls, Archibald accompanied her on long treks. The early spring cornfields were embroidered with green shoots amidst old corn stubble. The crunch of boots flattening the tiny ridges of soil forced up by the permafrost provided a fast-paced rhythm as they sauntered, and Tex sometimes had to flap her wings to keep pace. She often emitted a guttural purr from deep in her throat. They would walk for miles. When Tex got thirsty or tired, she led George to the water bucket outside the shack.

On May, 1982, Tex laid another egg. George put the egg in the next of an experienced crane for brooding, while Tex's nest was refilled with a mock egg. Tex would sit on the imitation for a while, then arch her neck and purr—signaling Archibald to take over. He dragged his sleeping bag outside the shed and gently covered the egg, set up a card table, and straddled the nest with a chair. There he would read or write, while Tex, temporarily relieved of her duties, went off to forage alone. Although sometimes out of her sight, Archibald communicated with Tex by listening to her calls. A low purring meant "You're here, I'm here, everything's okay." If Tex lost track of George when she wandered too far, she would give a high, plaintive location call, to which George responded, "Tex! Tex!" A shrill staccato call from Tex signaled alarm—a skunk, for example—and George would jump up and rush to her aid. If ever she glanced back across the field and caught George standing up by the nest, she would rush back to take over. They shared parental duties in the manner of wild cranes.

Sometimes wild sandhill cranes, which summered in the marshes of southcentral Wisconsin, alighted in Tex's territory. A normal crane would have attacked them. Tex did not even recognize them as cranes, let alone as intruders. She gazed curiously at the winged curiosities, then continued foraging for earthworms and grasshoppers. But if another human—a secretary bringing office mail to Archibald or an assistant with a supply of food, for example—trespassed, she attacked, and George had to go to the defense of the messenger.

Although tests proved the egg fertile, it had such a thin shell that it quickly began drying out. Archibald figured the only hope was to risk dipping the egg in a bucket of very cold water for five minutes a day. This would decrease the pressure inside, and draw liquid in. Though Archibald feared the chick would die of cold shock, on June 1, a weak, dehydrated crane—not unlike Tex as a hatchling—broke through the shell. It required intensive care, including tube feeding. But it survived. Archibald named it Gee Whiz, after George Gee, who had developed the technique for artificially inseminating cranes. Archibald's successful courtship with Tex, meanwhile, was soon reported by *Sports Illustrated, Time,* and *The New Yorker.*

On June 22, three weeks after Gee Whiz's birth, Tex was killed by raccoons. A late spring frost that year had made berries scarce in that part of Wisconsin, so the raccoons, which depended on them for food, took to eating chickens from neighboring farms.

When the farmers locked up their chickens, the hungry raccoons turned on the birds at the International Crane Foundation. Prevented from fleeing or defending herself by the closed quarters of the pen where she slept at night, Tex was bitten in the neck. Four other cranes died in the massacre.

Archibald had been scheduled to appear on the Johnny Carson show that evening to talk about his courtship with Tex. Word of the tragedy reached him at his hotel in Los Angeles moments before he left for the NBC studios in Burbank. He broke the news that night on national television. Over the next several weeks Archibald received numerous letters of sympathy. Tex's obituary appeared in *Newsweek:*

> Tex, the 5-foot female whooping crane that last month produced its first chick after years of ritual mating dances with a scientist; following an attack by a raccoon in its pen at the International Crane Foundation, in Baraboo, Wis., June 23. A member of an endangered species, Tex was "courted" every year during its six-week breeding season by Dr. George Archibald, the foundation's director, in an attempt to get it to ovulate. On June 1 the artificially inseminated crane gave birth to a male chick named Gee Whiz, which survives.

Today George Archibald speaks fondly of Tex, but always keeps a wall of scientific detachment that hides his true feelings. He has several pictures of her in his office in Baraboo. On the wall above his desk is a large colorful portrait of a whooping crane, crafted out of various types of beans and seeds. Only Archibald and a few others know it is Tex.

The story of Tex and Archibald has become a modern chapter in the lore of one of this country's most graceful birds, known for its

annual migration from northern Canada to wintering grounds in
southern Texas, the beauty and grace of its six-foot wingspans,
and, perhaps above all, for its habit of often remaining paired for
life. They are among the few animals that form lifelong attach-
ments—outdone, perhaps only by wild geese, some of which are
actually betrothed in the spring following their birth, a full year
before coming of breeding age. Engaged in adolescence, wild geese
often remain paired for life. In fact, among the most striking traits
of birds in general is that over ninety percent of the 8,600 species
are monogamous. While a few practice "permanent" monogamy,
the majority are "seasonally monogamous," remaining together
through the breeding season.

Monogamy among birds is, of course, a biological, not a
moral, precept. Cranes were probably practicing it for millions of
years before humans invented the idea that a man, in the phrasing
of *Genesis*, should "cleave" only to his wife. For some humans—in
many Western cultures, at least—the idea of monogamy has become a vehicle of morality; for birds it is a matter of survival.

For them, the demands of brooding and protecting the eggs,
and then raising the chicks, are so severe that it is usually impossible for a single parent to accomplish the task. To begin with, the
female makes a tremendous investment of energy before the egg is
even laid. An egg may weigh up to twenty or even thirty percent of
her weight; a whole clutch may exceed her total body weight. After
the eggs are laid, it is often incumbent upon the male to assume a
proportionate share of the responsibility for them. Eggs require
almost constant brooding, sometimes for weeks, and clearly it
would be difficult to do without a sharing of duties. In only a few
instances is one parent able to brood the eggs alone. The emperor
penguin is one such bird. While the female goes off, the male
remains at the nest for up to sixty-four days, fasting all the while.

While a number of different factors are responsible for the
monogamous way of birds, one of the basic reasons is the idea of

parental investment. Elaborated by Robert Trivers in an important 1972 paper titled "Parental Investment and Sexual Selection," the theory went a long way in explaining why cranes and most other birds favor monogamy. In short, the more each parent must invest to ensure survival of offspring (and hence, its own genes), the less it can afford to desert. How much parental investment is required by parents can depend on many things, the environment being central among them.

The breeding ground of the whooping crane is Wood Buffalo National Park in Canada's Northwest Territories. It is an extreme region, consisting largely of small ponds and shallow lakes separated by ridges covered in dwarf birch, willows, black spruce, and tamaracks. To assure themselves of adequate food in such a marginal environment, cranes must protect enormous territories—usually about 7.2 square kilometers, but sometimes ranging up to 34.8. Like someone trying to run a large estate single-handedly, a crane would be hard-pressed to protect such a territory alone. Added to this is the task of scouring the territory to gather sufficient food for the chicks. The naiads of dragonflies, larvae of caddis flies, mayflies, and other insects that make up the bulk of their food are scattered and sporadic, so that gathering them is a time-consuming task for parents. Furthermore, there is the responsibility of protecting the nest. Were it not for two birds watching the nest, predation by lynx, black bears, and wolves would undoubtedly be high. The moral of monogamy in this case is straightforward: two pairs of eyes are better than one. While the male generally guards and the female tends the nest, they sometimes alternate duties, which may include such passive defense measures as border patrol and more active ones such as luring a predator away from the nest by feigning an injured wing, while the other parent remains with the eggs or the chicks.

It is not only through the sharing of labor that monogamy helps pairs meet the rigors of raising a brood. By remaining together year round, and hence foregoing the time-consuming rituals of acquaintance every spring, it gives them a jump on the short northern breeding season. Paired cranes actually begin their courtship dancing in January in their Texas wintering grounds at Aransas National Wildlife Preserve. By the time they swoop down over the evergreen and into Wood Buffalo National Park in late April, they are nearly ready to nest, mate, and lay eggs. They lay two brown-speckled eggs in early May, with chicks breaking free a month later, in June. But by mid-September, summer's window begins to close, with mirrored lake surfaces turning white with ice and the swirls of snow scattered by the blowing boughs of the tamaracks. Within the period of eight weeks, a fledgling must be transformed into an elegant, four-foot-tall bird capable of an arduous 2,700-mile journey southward, the length of Canada and the United States, to the warm Texas wintering grounds. There is little time to spare, and the enduring pair-bonds of the cranes enable them to make the most of the brief season.

Studies of the cranes have yet to measure the precise benefits of remaining together year round, but if the kittiwake gulls in Britain are any indication, the savings are critical. It was found among the gulls that those who retained their mates from the previous season laid eggs up to a week sooner, produced a higher seasonal total, and reared a larger number of chicks than did those who took new mates. Among cranes, there is even the added benefit of having a mate to help them secure large territories in Texas as well, for whoopers are one of the few birds that remain territorial outside of breeding season.

Just how crucial a role a demanding environment plays in monogamy is shown by that fact that monogamy and polygamy often follow the lay of the land. Of the 291 perching songbirds (passerines) in North America, for example, 277 are more or less monogamous. Not surprisingly, thirteen out of the unfaithful fourteen breed in the marshes. Here, the sun concentrates its production of food in a shallow layer of the earth's surface, making food abundant and accessible, while the open spaces make predators easier to spot. But the food of these areas is often concentrated in patches. The male who owns a rich patch often attracts several females, who would rather share a male who owns a rich territory than have exclusive claim to a male in a poorer one. Once in a good territory, a female may be able to raise a brood unassisted. In short, the females don't fall for a male so much as for his territory. However, move from marsh to woodland, where a more equal distribution of resources prevents a male from monopolizing them, and the species there are more likely to be monogamous.

The term "monogamy" comes from the Greek *mono*—one—and *gamos*—marriage. If either sex takes more than one mate, it is called "polygamy"—many marriages. The area in between, like the colors of a spectrum, contain many blends. The crane and wild goose are at one extreme. Most other birds are considerably less faithful. The seasonal monogamists, such as the Eastern bluebird, almost always change mates each breeding season. Other species change mates with each brood. Still others strive for the best of both worlds. The great blue herons and great egrets of Hog Island, Texas, for example, will mate with one female but court other females in case their first pairing falls through—a sort of mate insurance policy. Some species that are usually monogamous, such as some swans, doves, finches, titmice, wrens, warblers, thrushes, and flycatchers, sometimes opt for polygamy. In one of the most intriguing cases of flexibility, the pied flycatcher pursues what is euphemistically referred to as a "mixed reproductive strategy."

In the spring and summer of 1979, zoologists from Uppsala University in Sweden set out to document the sexual strategy of the pied flycatcher by setting out 450 nesting boxes in the woods around Uppsala, tagging the birds who nested there, and keeping their activities and whereabouts under constant surveillance. What they discovered was that while nearly seventy percent of males in one population were faithful to one mate, the other thirty percent mated with one female, remained with her until she laid her eggs, and in the time between laying and hatching, flew to a distant territory, and mated with a second female. The males then returned to their "primary" females to help them rear their broods. Their "secondary" females were left to fend for themselves. This is better known as cheating.

Not surprisingly, the secondary females were able to produce far fewer young than the monogamous ones. Why, then, did the females not desert the cheating males and find mates who would devote themselves to their offsprings' survival? Because a cheating male apparently deceives his two females by keeping yet another bird's territory between them, so he will not be discovered. He is careful to remain around until his secondary female has laid her eggs and, in effect, is in too deep to back out of her commitment to the brood, even after she is left alone. If she were to discover his unfaithfulness immediately after mating but before laying eggs, it would still be worth the effort for her to find a new mate. But once she begins laying, she stands to gain the most by remaining with the eggs, with or without a mate, even if not all of the brood does survive. The males have evolved to capitalize upon the fact that the season's first brood fares better than later ones, largely because food supplies peak early and almost immediately begin to wane. Therefore, the female must balance the promise of this first batch against the declining fortunes of a later one.

From the male point of view, then, once the eggs have been laid his task has been accomplished. Even if the truth were known, the female would still be forced to remain with the eggs. All the male must do is somehow delay her discovery until she has laid them. This way, she is likely to remain and attempt to rear the brood. His genes are propagated at minimal expense to himself—but at a heavy price to the female.

Yet another factor enables the deceived to make the best of a bad situation. If the female loses her mate at laying time, she compensates for the increased burden to herself by reducing the number of eggs. This suggests that at some point she "realizes" she is being cheated on. How does she find out? It is actually in the male's selfish interest to "tell" her once she has laid the eggs. By allowing her to reduce the size of the brood, more of his offspring will survive. There is a well-defined line marking the point of no return for the slighted female, and so the male must divulge the information after she has committed herself to remaining but before it is too late to reduce the brood size. Perhaps he conveys this information by simply refusing to incubate the eggs.

The situation of the male and female pied flycatcher lies at the center of conflicting reproductive interests of male and female: she can gain most by remaining faithful and encouraging the male to help raise the offspring. He, on the other hand, if there's a good chance the female can go it alone, gains more genetically by mating with other females, and thereby leaving behind more offspring.

A rare occurrence among birds, cheating benefits the pied flycatchers because the pairs defend not large surrounding territories but only nests—a job the female is capable of doing alone, if need be. Where a larger territory has to be defended, as among cranes, the male's absence would probably spell disaster for the brood. The female pied flycatcher seems to have evolved none of the countermeasures against male infidelity that many other birds have. The females of some species force their mates into long and costly courtship performances, pushing them past the point of no return in their relationships—that is, to the point where the genetic expenses of desertion are greater than those of remaining.

The male can recover his initial investment only by staying around.

Courtship may also be a way of testing a male's true availability. Niko Tinbergen once saw a female red-necked phalarope vigorously court an already committed male who had taken leave of his eggs to feed at a nearby pond. Whenever he tried to copulate with the flirtatious female, she fled—a coy performance carried on for several days. Tinbergen suspected that somehow the female knew the male had a prior commitment, and didn't want to risk her own eggs to an encumbered male. To test his hypothesis, Tinbergen destroyed the male's eggs. When the male again went to the pond and approached the soliciting female, she almost immediately copulated with him. How did she know that he was suddenly free to invest exclusively in her own brood? It was probably through courtship, which tested just how far he was willing to follow her. Following her a substantial distance must have meant that he was not preoccupied with another brood hidden somewhere nearby.

From even a scientific perspective, there may be some foundation to the dove's role as a symbol of peace. Monogamous, it is freed from some of the competitive conflicts of highly polygamous animals. The red-crowned crane's stalwart, monogamous qualities have made it a symbol of long and happy marriage in Japan, while in India the Sarus crane is a symbol of fidelity and true love. There is often a scientific lining to the romantic claims made about animals, not least of all about cranes, which are said to sometimes pine to death if separated. In fact, though cranes will readily adopt new mates if an old one dies, given the opportunity they may remain together for decades.

George Archibald tells the story of a wild sandhill crane he watched one summer on a roadside near Baraboo. "It stood there like a gray ghost, from dawn to dusk, every day from June till October. A state trooper told me its mate had been struck and killed by a car as the pair crossed the road. But their bond was so strong that when the female died, her mate returned to the same place every day, hoping that if he just waited long enough she would eventually come back to life."

10.

The Price

of Paradise

Just as I got home, I overtook Ali returning from shooting. . . .
He seemed much pleased, and said, 'Look here, sir, what a curious bird.'

—Alfred Russel Wallace,
The Malay Archipelago

IN 1522, when the *Victoria*, the only surviving ship of Magellan's around-the-world voyage, sailed into the harbor at Seville, among the curiosities on board were the skins of two colorful birds. In the process of skinning and preserving them, native Papuans had removed the birds' feet and legs, so in the minds of the first Europeans who saw them, the skins bespoke of svelte, legless, wayward passengers from paradise. Footless, they must have remained airborne all their lives, the amazed Europeans reasoned, abruptly visiting the ground only at the moment of death. This would have entailed novel arrangements for breeding: after mating in mid-air, the female would have had to lay and brood her egg in a groove

130

between the male's wings while riding piggyback. Thus reared, the chick presumably carried on the species' tradition of never touching the ground. For food, these fantastic creatures probably subsisted on dew, perhaps gleaning it from the air or from the broad leaves of the lush tropical forests where they lived.

For all this detail, it wasn't until 1824 that a live specimen was first seen by a white man—a ship's apothecary. Neither footless nor bound to a diet of dew, the mysterious bird actually lived on berries, fruit, and insects, and spent most of its life firmly anchored to branches with quite massive feet. Nonetheless, to this day the greater bird of paradise still bears the name *Paradisaea apoda,* or footless one from paradise.

Native only to New Guinea and a few outlying islands, and to parts of Australia, these lavish birds are most closely related to the crow. But while some species do indeed resemble their plebian American relatives, most sport far more lavish plumage and practice elaborate courtship rituals. As many as forty copiously plumed males may gather in one tree and simultaneously display before females. Charles Darwin believed such plumage gave the males "power to charm the females," though biologists still puzzle over its undoubtedly wider significance. One male's plumage, for example, may even serve to dominate a lesser-endowed male.

Outside of New Guinea, one of the world's largest concentrations of the birds is along the western edge of Central Park in New York City. Unfortunately, this flock is stuffed and tucked away in gray specimen trays on the fourth floor of the American Museum of Natural History. All forty-two species are represented in the collection. One of the largest, the greater bird of paradise, is about the size of a raven; one of the smallest, the king bird of paradise, is about six inches long including its short tail. The king bird of paradise's head, throat, and entire upper body are crimson, which crescendoes to an orangish glow on the forehead, where its feathers extend beyond its nostrils and halfway down its beak. The breast and belly are silky white, separated from the reddish upper body by a bold metallic green band. The lesser superb bird of paradise, while more conservative in color, is even more brilliant in overall effect, with velvety black plumage and a dusting of brass on its neck. Its chest has a shield of satiny bluish-green feathers. But its most exotic feature is a black fan-shaped cape, which during courtship is spread vertically behind the head and accents the turquoise shield. Then there's the red bird of paradise, the several species of six-wired birds of paradise (so-called because of the growth of wire-like quills from its head, streaming back with spreading, fan-shaped tips), and the paramount Twelve-wired bird of paradise, with its two-inch beak and half dozen "wires" curling back from each flank under its violet wings and beside its tail. The beauty of these old specimens is astonishing; a hundred years tucked away on shelves has done little to subdue the iridescent splendor of their plumage.

The collection is presided over by Mary LeCroy, Senior Scientific Assistant at the museum, an authority on birds of paradise and a protégé of renowned ornithologist Thomas Gilliard, who spent much of his life studying the birds in New Guinea. LeCroy points out that this collective overview of the family makes it boldly apparent that, with few exceptions, the brilliant plumage of the males eclipses the drab feathers of the females. It is as if the latter were juveniles or even a different species altogether. As if to make the point, LeCroy holds up a greater bird of paradise female in the light of a window overlooking Central Park. Its wings and back are brownish, with a slightly lighter brown belly.

"Sexual dimorphism [the contrast between male and female] in the genus *Paradisaea* is as extreme as in any genus I know," says LeCroy. "You have a profusion of male plumage that has assumed every color of the rainbow and has grown to improbable lengths. The females, on the other hand, seem to have evolved in a

132

totally opposite direction—becoming drably colored with simple plumage."

LeCroy points out that the extreme sexual dimorphism among many birds of paradise is closely tied to the fact that they are among the small percentage of birds that are polygamous. Generally speaking, species in which the male and female are similar are likely to be monogamous. Indeed, if you look at the twelve species of birds of paradise that are thought to be monogamous, the males and females are remarkably similar, while in the thirty-odd polygamous species, the males and females are very different.

In monogamous societies, one-to-one pairing means that theoretically most animals can find a mate without excessive competition. But in polygamous societies one male may mate with many females, meaning most males are left without mates, and males must compete very hard to be among those who achieve the chance to produce offspring. LeCroy carefully replaces the greater bird of paradise in the tray and holds another female specimen in the light, this time an emperor bird of paradise. With a brownish rump and wings fading to a plain yellow at the crown and finally to brown at the forehead, it is nondescript compared to the male.

Some biologists believe that the extravagant appearance of males in such societies is the direct result of pressure to "please" families. To put it simply, if the female's preference is based on color, then males over time will become brighter; if it is based on length of plumage, males will develop long tails. Edward O. Wilson describes the ardent efforts of males to win females as "a contest between salesmanship and sale resistance."

The analogy might serve as well to explain the sort of "design" pressure males are under. If customers find a certain design feature of a particular make of automobile highly desirable—a racing stripe, for example—within a short time many of the cars appealing to that class of customers will be "selected for." As each manufacturer attempts to outdo the competition, it's likely that the racing stripes will grow bolder each year. What began as a marketing experiment (that is, an evolutionary quirk) may become standard equipment, with stripes growing more exaggerated with each generation of cars. While evolution hasn't the benefit of foresight, the analogy does reflect traditional theory on why male birds of paradise have developed such elaborate plumage.

However, other biologists (Mary LeCroy among them) believe that it is not the pressure of female preference so much as the pressure from other males that lies behind such extravagant appearance. Among some species such as Goldie's bird of paradise, males appear to settle the competition for females among themselves by establishing dominance hierarchies based on the elaborateness of a male's plumage. When a female arrives at a group of displaying males, the dominant one steps forward, while the others defer. "Elaborate plumage, which has usually been considered a means of attracting females, may actually serve males in the contest for dominance," says LeCroy. "But whether it is the result of female choice, male competition, or both, males become more colorful."

Similarly, if dominance is based largely on size, then males will get larger. But whereas some large animals go in for physical warfare, the bird of paradise goes in for visual combat, so to speak. Feathers get longer, colors brighter. "What once was a drab, crowlike bird may, in the long course of evolution, become one of these": LeCroy picks up a male emperor bird of paradise, its brilliant emerald face and chest rippling in the light, white feathers spilling from the flanks like long strands of silk. "At the height of its courtship, it slips over on its perch like this," she said, spiriting the bird upside down to reveal its acrobatics.

Among birds of paradise, the most flamboyant "salesmen" are found among the seven species that use what is known as an arena display—a designated area of the forest where males gather, call loudly, then display for females. One of these species, the magnificent bird of paradise, chooses a slender tree on a steep hillside, then strips the tree of leaves until shafts of sunlight shine on his display area. He calls attention to his own court by emitting loud *eeeks* or *eees,* intermittently picking up twigs in his beak and dumping them outside the cleared area. When a female arrives, usually pausing on a nearby branch, he splays his iridescent shield of breast feathers—a rich green overlaid in hues of purple—and ripples it in the sunlight. The velvety brown head feathers, extending partway down the beak, bury the nostrils. The feathers surrounding the eyes are raised slightly, revealing an iridescent spot in front of each eye. If the female appears ready to abandon him, he may jump from the perch and quickly begin clearing the ground of any remaining twigs, leaves, or bark.

After this routine is repeated several times, the female perches above the male, and both hang sideways on a nearly vertical sapling. Slowly she may move down toward him. The male extends his body almost horizontally from the perch by gripping tightly with his feet and leaning backward. He extends a yellow feather cape from his nape like a radiant halo as his tail appears to vibrate with the strain of his awkward position. As the female moves closer, he abandons his strenuous position and mates with her, then falls into yet another display: he lifts his tail at a right angle to display its glittering surface. The two wiry, steely-blue, ten-inch-long feathers, curving outward like the handlebars of a mustache, with the tips of each curving in to form two double circles, are fully revealed. As part of the post-copulatory display, the male pecks at the female's nape, pausing each time to "gape" and reveal the intense yellow-green lining of his mouth. When the female, who meanwhile is garbed in drab browns, departs, the male flies about the display area, finally settling on a branch and calling loudly.

Interestingly, differences in the appearance of the males and females seems to be echoed in their behavior. The male's role in courtship, while certainly no more important than the female's, is—like his plumage—considerably more extravagant. But whatever the female may lack in plumage (at least as measured by human standards), her appearance serves her well. Instead of pouring energy into the development of elaborate costumes or outlandish displays, females invest it, perhaps rather more sensibly, in parenting. The female's drab appearance makes her less visible to predators. While males may be selected for elaborate plumage, females are selected for camouflaged plumage that allows them to remain undetected at the nest—in other words, for traits that add up to good parenting.

It is said that ostentation sometimes precedes extinction. In the case of birds of paradise, there is little doubt that one way or another, the male does pay a steep premium for his appearance. The encumbrance of feathers, for example, makes escape difficult.

While among birds of paradise the liabilities of such plumage may be minimal, since predators are few, they are more evident in another generously endowed species, the great-tailed grackle of southcentral Texas. The male's large tail, while ideal for attracting females, makes it difficult for him to fly in strong winds. The less-endowed females, however, can fly more easily. It is not surprising that during winter, male grackles have a death rate twice as high as females—probably due in large part to their need for more food than females to maintain their larger, more elaborately plumed bodies.

Conspicuous males of some species, however, have evolved ways of coping with the liabilities of notoriety. One way is to limit colorfulness to the breeding season. The male African paradise widowbird, for example, wears a pair of tail plumes several times his own length during breeding season, but loses them afterward, leaving him looking very much like a female. Shedding such accessories not only makes the male safer from predators, it makes socializing with other males easier, for the very thing that makes such males attractive to females often also makes them threatening to other males—as a result of which cooperative ventures such as flocking are almost impossible.

Some males are equipped like traveling magicians, able to pack up props at a moment's notice. The male magnificent frigatebird swells a great scarlet bladder beneath its throat during courtship, allowing it to collapse afterward. The three-wattled bell bird displays three spiked wattles at the base of his beak. The ocellated turkey raises his tail and breast feathers. The great bustard can transform himself from an inconspicuous creature almost unnoticeable on the grassy steppes to a vibrant figure visible to females hundreds of meters away, only to fade again into the landscape once his courtship has been accomplished. Some birds have a striking color hidden in the lining of their mouths that can be seen only when they are "gaping" at a female. All of these strategies give such birds the benefit of appeal without year-round liability.

The bowerbirds of Australia and New Guinea may have evolved the most original solution to the problems associated with attracting females. Instead of wearing attractive plumage, they build elaborate courtship bowers, decorating them with a multitude of glittering objects—from flowers to discarded cigarette wrappers. The Vogelkop Gardner bowerbird, for example, lacking a crest and indistinguishable from the female, builds an elaborate teepee-shaped bower on the ground and decorates the ground around the entrance with a pile of brightly colored items collected in the forest. A spotted bowerbird was once found to have fragments of glass, shell casings, and even a toothbrush. Thomas Gilliard hypothesized that by transferring his attractiveness from himself to his bower, the male bowerbird can attract females without the year-round hazard of conspicuous plumage. Further evidence of this "transferal effect" comes from the observation that, within a single genus, the less conspicuous the species of bower-bird, the more elaborate the bower built by the male.

The birds of paradise, however, must live with the liability of raffish plumage, and no predator has made them pay a higher price for their beauty than man. While native Papuans traditionally used the birds' feathers for ceremonial dress, it was the Europeans who posed the first serious threat to the birds. During the first five years of German colonial rule in northeastern New Guinea, more than 50,000 skins were exported to feed the millinery markets of London, Paris, and other cities. Between 1904 and 1908, in only eight sales, at least 155,000 skins were auctioned off in London alone. Not until 1924 was commercial collecting banned. But by then European influence and the breakdown of native traditions further threatened several species.

The trees where the birds displayed were traditionally owned and protected by the individual on whose property they were located. Only plumed males were killed—usually knocked from trees by blunt-tipped arrows so as not to draw blood and damage the feathers. With the birds performing in the same arenas through generations, and replacement birds available from young males, a measure of protection was afforded. Ironically, frequent warfare among the tribes may actually have benefited the birds by making much of their habitat a no-man's-land—a vacuum that was rapidly filled by animals.

With the arrival of European law and order, however, tribal conflict decreased. Natives could roam freely—and poach birds from trees owned by others. Furthermore, the growing tourist trade to the once isolated islands sparked a certain entrepreneurial spirit among the natives, setting their ceremonial dress on an escalating course of evolution similar to that of the birds themselves: costumes got bigger and more lavish to please the new guests. In the 1960s it was estimated that at the annual Mount Hagen show in the central highlands, dancers wore bird of paradise feathers worth some quarter of a million dollars. The complex ecological web, built largely around the birds' spectacular appearance, had, indeed, begun to unravel.

Far and away the greatest current threat to the birds is the destruction of their habitat. Despite the heavy pressure direct exploitation once placed on the birds, they have continued to flourish where their habitat has remained intact. But in many areas this link is being severed as forests are cut and increasing population pressure prevents their regrowth. Slash and burn agriculture is eradicating the forests at an increasing rate. And in recent years outside interests have been clear-cutting forests for timber, a practice from which the forests are unable to recover. As goes paradise, so go the birds.

11.
Why Mammals Are Unfaithful

THE GUINEA PIG is the only mammal sufficiently well developed at birth to survive without milk. At the other extreme is the pilot whale calf, which may suckle its mother for up to seventeen years. Whatever the nursing period, for many mammals the breast is among the most essential bonds between mother and young. Searching, puckering, zeroing in on its scent, infants universally rally around it. For humans and other mammals alike, weaning is a traumatic rite of passage, the mother resolutely pulling away, her frantic infant curling its lips in a final, desperate pucker.

The breast surfaces at many levels of the mother-infant bond, including nourishment, protection, security, and warmth—a bond that indeed runs deep, in many animals as well as in humans. Jane Goodall tells the story of the matriarchal chimpanzee Flo, mother of numerous young. After her death one of her youngest offspring, an adolescent son called Flint, was deeply shaken, and consoled himself by visiting the nests and resting places they had shared in the rain forest. "For Flint it was like losing his whole world," Goodall recalled. Three weeks after his mother's death, Flint, in a state of grieving, also died.

Not all milking stations have taken the literal form of breasts, or even nipples. Some mammary glands closely resemble the sweat glands from which they may have developed—tiny surface springs in which newborns bathe to keep moist. Female duckbills, or platypuses, primitive egg-laying monotremes of Australia, have lapping stations on their stomachs, where milk escapes from pores and the young gather to drink like desert beasts at a favorite watering hole. A North American marsupial, the opossum, gives birth to honeybee-sized embryonic babies, which migrate across the furry terrain between the vaginal gateway and the pouch—where survivors of the journey latch onto tiny teats. But while the American opossum produces as many as eighteen embryos at once, it has room at the nipples for only seven. Those who fail to qualify in this first elimination heat, perish. Sperm whale cows have nipples, but their young lack maneuverable lips and cannot suck. After birth they lie on the surface near their mothers, blowholes above the surface, and while the calf places its open mouth against a nipple, the mother pumps milk.

The influence of the breast, especially in the higher mammals, reaches far beyond a newborn's life. Among humans, who possess the only mammary glands that remain highly conspicuous even when not producing milk, the breasts may serve as ever visible sexual stimuli that aid bonding. They may also serve as signals of appeasement. In *The Iliad,* Helen placates her angry husband by baring her breasts, while in some cultures mothers pacify their children, even those well beyond nursing age, by allowing them to fondle their breasts. (In Western societies such functions have been largely obscured by the portable teat, or baby bottle.)

Motherhood can be divided into three stages. First there is the period of preparing for birth, which may include building a nest. In some species this may involve building several nests; raccoons, for example, give birth up in a tree, then move the young to a nest on the ground. Then there is the time of nursing. Finally there is the period of weaning, when the young animal must learn to hunt for itself. Weaning is often a stressful period, especially among predators, for there is usually a long period in which the offspring are too old to suck, but too young and weak to hunt and kill on their own. They must depend on the mother for food, while at the same time learning to hunt for themselves. Some mothers will bring back to the nest or den animals that are wounded but still alive, leaving them for the offspring to practice on. Lion cubs might accompany their mother on hunts for a year before actually bringing down an animal themselves. Among social animals, the infants might become part of a herd or troop, while among animals who live alone, the offspring are simply abandoned one day. In this rite of passage for grizzly bear cubs, the parent chases them up a tree—as it may have done previously in times of danger—but this time it leaves them behind.

The female Australian koala, a herbivore, creates a transitional diet of pureed eucalyptus leaves during weaning, processing them in a special organ, then passing them through her digestive tract. The processing occurs on a precise daily timetable so that the food remains safely separate from the mother's excrement.

This kind of close-knit relationship between mother and infant obviously has great implications for the father, who may be fundamentally excluded. He is forced to look for alternative ways to spend his time and energy. The less opportunity he has to invest meaningfully in his present offspring, the more sense it makes for him to defect and produce another brood elsewhere, eventually abandoning them as well. This polygamous behavior maximizes his genetic investments. There is a degree of mathematical predictability in the choice he makes: the higher the female's investment, the greater the male's temptation to desert. But there is another complicating variable in the equation: the less chance a male has of finding a second receptive female, the more it makes sense for him to try to assure the survival of his offspring from the first. He must measure opportunity against risk. But all things considered, the domination of parenting by the female, symbolized by the exclusive claim to the breast, often makes polygamy a worthwhile alternative for male mammals—worthwhile, that is, in the sense that this route will allow for the survival of the maximum number of his offspring. It is little wonder that almost all birds are monogamous, but that ninety-seven percent of all mammals are polygamous.

Where the opportunity for meaningful male participation in rearing young is available, one finds a different story. The monogamous male fox plays a vital role in catching food in a protein-poor environment. Among wolves, which are also monogamous, the male may help feed and protect the young, as well as provide opportunities for learning. Beavers, who sometimes remain together for life, divide the tasks of maintaining a home and parenting between the sexes. One of the most monogamous of mammals is a small antelope known as the klipspringer—Afrikaans for "cliffhopper." Klipspringers are very highstrung, for predators are everywhere. The only recourse for the klipspringer is to team up with another klipspringer, one to act as the eyes and ears for the other while it grazes. One is automatically bound to the other. As it happens, the male is probably better at detecting enemies than the female—just compensation for the female's large investment in having to carry the young until birth. Without each other's assistance, few klipspringers would survive. Genetically speaking, the male has little recourse but to invest his energies in one female rather than leading a promiscuous life.

This is not to say that the polygamous male, in being freed from parental duties, somehow "wins" on all counts. It's simply that rather than pouring energy into parenting, he invests it in producing more offspring. This may entail the burden of protecting a group of females, or a harem. Often it may also involve stiff competition with other males.

143

Probably nowhere is the polygamist's burden more demanding than among male elephant seals. Like all seals, they must forego the water—an environment they are designed for—and drag their rotund bodies onto a beach to breed. The males haul themselves out of the water in December and stretch out on beaches in scattered pockets from Isla Cedros in Baja California, Mexico, to the Farallon Islands in central California. At only one other time, during molt, do they come ashore.

On land, elephant seals are peculiar in both habit and appearance. The size of a cow and weighing up to 650 kilograms, they are largely a cargo of fat hauled on land by two large front flippers. Elephant seal bulls are characterized by a long trunklike nose—a feature from which they derive their name. An adult female weighs less than half of a large bull and, by comparison, might be mistaken for an infant. But while ill-equipped for land, their massive, blimplike bodies are transformed by the buoyancy of salt water and the creatures become animals of tremendous strength, capable of diving deeply and traveling as far as 3,000 miles from their breeding beaches.

The lives of dominant bulls, like those of athletes, are characterized by transient glory. In his prime, one male may claim a harem of fifty females. Not only will he go for three months without eating in order to devote his entire efforts to copulating, but he will even abandon one of his own matings in order to prevent another bull from copulating.

Once they arrive, the males at each beach engage in tests of strength and will. Deadly confrontations are usually prevented by a kind of ritualized saber rattling in which a male raises his forequarters eight feet into the air, tilts his elongated head over backward, and dangles his fleshy proboscis into his gaping mouth, while emitting loud, low-pitched guttural warnings. If another male returns the threats, the confrontation will escalate into combat. One bull will slam its chest into the other, sumo style, with each gripping the neck of the other with his long canines. Wads of protective fat may be ripped away, resulting in profuse bleeding of the wounds. If several pairs of males simultaneously engage in battle, the tide line along a stretch of beach may go red with blood. The fight, usually lasting no more than fifteen minutes, ends when one of the fighters turns and flees. In this way, social rank is established. Males of descending rank fan out from the alpha male, establishing a rigid though frequently changing social order, remarkably similar to the pecking order of domestic chickens.

Given the size and strength of elephant seals, their territories are surprisingly small—often less than a hundred and twenty square yards—compared to many mammals. This is probably because their blimpish bodies are not mobile enough to defend larger areas.

Females begin hauling out of the water in mid-December, several weeks after the males. Their first two priorities are to give birth and to wean pups who were conceived the previous December and carried during the year at sea. Within a week the scenes of bloody male combat give way to a more maternal setting in which hundreds of tiny newborns suckle their mothers, who have gathered within the territories.

Even before the one-month weaning period is finished, the bulls begin clamoring to mate. Always initiated by the male, the mating sequence begins as he approaches the female, laps a front flipper over her back, and rolls her small body to its side so they are belly to belly. If she doesn't oblige, he chases her and plops the full weight of his head and neck on her back. A resisting female swishes sand back in the face of the male with her flailing flipper while emitting a croaking call, then swings her hindquarters around and smacks her back flippers solidly against the male's extruded penis. While a male sometimes will continue to pin a resisting female or clench the back of her neck with his teeth, without her cooperation there is little chance of mating—leaving the female with a larger say in mating than might readily be apparent.

For females, life is an unbroken circle of mating, gestation, and weaning their offspring. During mating, the aggression of the males increases. As they charge across the sand in pursuit of females or other males, pups are frequently crushed and die of ruptured organs and internal hemorrhaging. The deep bellows of males, the alto groans of females, and the shrill cries of pups form a three-part chorus that reflects three distinct levels of pursuit: competing males, protesting females, and pups caught in between. Newborns less than a week old are usually victimized. In fact, nearly half of the pup deaths in a single season can be attributed to adult males. It may qualify as a form of infanticide—one male's killing of a competitor's offspring.

The elephant seal presents an accessible and fascinating subject to study, although not without hazards. Burney Le Boeuf, a biologist at the University of California, Santa Cruz, has been studying the seals on Año Nuevo Island in San Mateo County, for eighteen years. One of his early problems was identifying individual animals. He finally solved this by filling a plastic spray bottle with a mixture of Lady Clairol hair-coloring and hydrogen peroxide, sneaking up beside a sleeping male, squirting a number or some initials on him, and then, with luck, escaping before it awoke. With disturbed males swinging around and lunging for him, he sometimes reverted to using a paint-pellet pistol from a safer distance.

Keeping a record of males within the population revealed some surprising results. "In one study conducted on Año Nuevo Island, four percent of the males mated with eighty-five percent of the females—an extreme degree of polygamy that few other animals approach," he says.

According to Le Boeuf, weaker bulls, unable to gain direct

access to females, often wait at the edge of the breeding ground and attempt to mate as females return to the water after mating, although most have probably already been fertilized during the prior four days of mating with the dominant bulls. Others may try to cross the perilous lines of defense around the receptive females, being met each step of the way by males of ascending rank. Still others may wait until the dominant male has just finished an exhausting fight, then initiate a challenge. This strategy has precipitated the fall of many alpha males.

Le Boeuf tells of yet another strategy he once witnessed in which the second-in-command, or beta bull (designated CLS), challenged the alpha bull, RNK. The challenger was roundly defeated in a long and bloody fight, and after a short rest, swam 300 yards around the island, hauled out near a different harem, and proceeded to threaten the resident alpha bull there. In this case, the alpha quickly abdicated, leaving CLS momentarily in charge. But CLS himself was soon dethroned by yet another high-ranking male, and had to settle for the number two, or beta, position for the rest of the season. Nonetheless, by switching locations, he was able to mate with more females than if he had remained in the shadow of RNK. The lesson is that where competition is too stiff, one might fare better by going elsewhere.

In comparison to the alpha male's conquests, the records of the second-, third-, and fourth-ranked bulls were unimpressive. During a single two-month reign, one alpha bull frightened off other males more than 700 times—almost as many as the beta, gamma, and delta bulls combined. The dominant male prevented other males from mounting more than 200 times, compared to a total of only 80 for the others. While having none of his own copulations interrupted, he cut short others' copulations more than a hundred times.

It is not surprising that many males never copulate at all, but are forced instead to sleep at the edge of the breeding areas. If they attempt to join the action, they are usually quickly chased back to their place of repose by a stronger male. This intense competition between bulls has selected for large powerful males, which in turn has created one of the trademarks of a highly polygamous society—males with exaggerated characteristics who can overpower other males by virtue of size. The relative size difference between male and female—one aspect of their sexual dimorphism—might be seen as a sort of polygamy index. Some biologists speculate that men, who are on average about ten percent larger than women, reflect the mildly polygamous behavior practiced by most humans.

The dominant elephant seal's success is not without its price. A male often spends himself in a single season. Exhausted and depleted of fat reserves, he may never again reign as harem master. Some alpha bulls take such a beating that they die right after the breeding season. Exceptional males may reign for as many as four consecutive years, but most usually die within a year or two of their best season.

The elephant seal is a good example of how a number of environmental characteristics have converged to create an exemplary polygamist. As sea creatures, both male and female seals have an abundance of evenly distributed food. Able to fish for herself, the female can easily supply her pup with ample milk. The large size of elephant seals may also contribute to their polygamous life-style. They tend to be large because large bodies retain heat better in the frigid water than do small ones. The elephant seal's impressive reserve of fat also enables him to fast during the three-month breeding season. Few animals are able to fast in order to devote energy almost exclusively to breeding. A small mammal the size of a shrew, for example, could sustain himself for only thirty minutes or an hour before having to pause to eat, at which time another male might move in on his females. Furthermore, since suitable breeding beaches are scarce, females tend to congregate in small areas—a standing invitation to be monopolized by a single male. This is known as resource-based polygamy.

Another case of resource-based polygamy is found among the South American vicuna, a diminutive member of the camel family. A male vicuna who claims a rare rich pocket of Andean grazing land may find as many as eighteen females coming to join him.

Polygamy can occur for many different reasons. When a male gains access to females by virtue of his ruling an area rich in resources—whether it be a suitable breeding beach or a source of food—he is said to be engaging in resource-defense polygamy. Lionesses, on the other hand, band together with a single male for mutual protection against other males. It has been argued that female zebras are attracted to the stallion who is skilled at fighting off the many predators that feed upon their young. In some animals, however, males offer neither defense of the females nor any particular resource. There is no trade-off of riches, nor any guarantee that the male will help raise the offspring. Such males gather merely to display themselves. The group display is known as a "lek"—a word thought to have come from the Swedish for "sport" or "play"—and the tree, pasture, or other area where they display is known as an "arena."

Although not uncommon among birds—birds of paradise, to name one family—lek polygamy is exceedingly rare among mammals, practiced only by a few species of antelopes and deer and by the so-called flying foxes, members of a group of huge, furry bats. With leathery wings spanning up to six feet, their bodies covered with fur, flying foxes appear, indeed, to be descendants of a happenstance mating between a California condor and a red fox. Of the nearly 1,000 bat species—comprising nearly a quarter of all mammals—nearly 200 qualify as flying foxes. While for various reasons these and other bats have acquired an unjustifiably shady reputation, they are, in fact, solid ecological citizens. Bats live everywhere except in extreme polar and desert regions and on a few isolated islands, but the flying foxes are limited to tropical regions of the Old World, where they play a major role in forest propagation. Indeed, their seed dispersal and pollination services are vital to rain forest survival. Many of the world's most economically important plants rely on flying foxes or other bats for survival in the wild. These include peaches, mangos, avocados, and bananas. Even the legendary giant baobab tree depends on bats for pollination.

Because bats are active at night and easily misunderstood, their reputation has been remorselessly maligned. In fact, bats are no more prone to rabies than raccoons or other wild animals. Gentle, intelligent, and sensitive, they are much friendlier than most. Far from being creatures from the ebb tide of evolution, recent neurological and morphological studies indicate that the flying foxes may actually be primates. Sadly, despite their numerous ecologic and economic values, many bat species, especially flying foxes, are declining at alarming rates and several are already extinct. Their loss may have serious, even disastrous environmental consequences.

The Gambian bat and the bizarre hammer-headed bat *Hypsignathus monstrosus*—roughly translated as "high-jawed monster"— of West Africa are noted for their leks. The hammer-head, common to the lowland rain forests of West and Central Africa but ranging eastward to Uganda, is the largest bat native to continental Africa. The male, nearly twice the weight of the female, sports a grotesquely swollen muzzle that flares out into lip flaps. Secretive, shy, and usually hidden under the dense forest canopy, the hammer-headed bat proved an elusive subject of study. Not until the 1970s did American zoologist Jack W. Bradbury—equipped with ropes and trapezes to hoist himself into trees, machetes to clear a pathway, and moon- and star-scopes to pierce the fading light of dusk—thoroughly document their extraordinary behavior.

For nearly four breeding seasons, or seventeen months, during which time he contended with Gabon vipers, aquatic cobras, and deadly black mambas, Bradbury studied the congregations at their arenas along the Ivindo River in Gabon. The forest canopy proved so inaccessible that he was forced to rig trees with fixed ropes and rock-climbing equipment, and hoist himself, click for clink, on ratchets designed for mountain climbers. Once he even attempted to use a crossbow to fire climbing ropes over the branches of the trees. In one serious mishap, the hardware jammed, leaving him stranded eighty feet above the ground, amid a swarm of angry, biting bees, in the westernmost reaches of Africa, twelve miles from the nearest town. The only way he could extricate himself was to untie himself completely from the rope system and stretch up to a slippery wet branch four feet away. After nearly losing his grip, he finally secured a desperate hold, untangled the rope, and slid back to the ground—but not before he permanently injured his back. Today, Bradbury happily teaches animal behavior at the University of California, San Diego.

Through eventually tagging, recording, photographing, and tirelessly observing the bats, Bradbury drew the most complete picture to date of their courtship behavior. Each evening during the months of June and July, up to 200 males fly in low over the rivers, sometimes skimming the surface for a drink. They take up residence on a small island where their favorite calling site is located. The typical calling roosts are horizontal branches beneath the dense umbrella of foliage. According to local residents, the bats had been using the same calling sites for as long as sixty years, or as long as anyone could remember. Once arriving at the site, the males warm up by rendering a raspy and harsh call, but as the pharyngeal pouch over each shoulder expands, the call soon takes on a clear metallic brilliance, like a "glass being rapped hard on a porcelain sink," as Bradbury described it. They also flap their leathery wings at about twice the rate of their calls. If challenged by another male, a caller will emit a series of aggravated gasps and honks as he lashes out with his wings. Sometimes the roosted bat surrenders his perch, only to reclaim it seconds later. Often a male claims several perching sites, calling from one for fifteen or twenty minutes, before switching to another. Only about a foot apart, each perch broadcasts his sounds slightly differently, and he is able to target females coming up from the south as well as the west. Occasionally two males will grapple, bite, and slap each other in flight. After settling their disputes, usually within a half hour of arriving, the males turn their attention to the females, who soon begin to arrive. When a male spots a female hovering nearby, he accelerates his calling and executes a rakish "staccato buzz" by pulling his wings tightly against his body and emitting several long buzz notes. If the female moves on, he resolutely extends his wings and assumes his regular calling, hoping the next female will be more impressed.

"It wasn't until our second year that we finally caught our first bat," Bradbury recalls. "The trick was rigging canopy nets at sixty or eighty feet in the middle of the aerial corridors in the canopy that the bats prefer for flight. Tagging the bats was a problem: we tried bleaching numbers in the fur, plastic bird rings mounted on the forearm bones of the wing, and tattooing the wing membranes. Finally we devised a method of using rubber cement to stick tiny three-gram radio transmitters on their backs. The first attempts were nearly disastrous. Already exhausted and shocked by being netted, the bats needed time to recover before flying off with the additional weight. When we tossed the first two radio-equipped bats into the air at the river's edge, they promptly fell into the water and began swimming. We had to jump into our dugout and rescue them. After drying them off, we let them hang on a high branch and leave at their own leisure. Letting them go when they were ready proved the key to success. The radios were the only method by which we could find bats at their day roosts, hidden in the forest canopy, and identify them regularly when displaying or foraging."

Bradbury learned that a female may visit five or six neighboring males in an evening, hovering over each in turn as if carefully comparing each to the other, before focusing her affections. As she lands beside him, the male suddenly stops calling and mounts her from behind. After about a minute, she signals the end of mating by loudly whining. She then flies away. The male pauses for a minute, then begins calling again in the contest for yet another female.

The lek, it appears, may be the ultimate in female choice. She is not forced into any particular lek, and once there, she is not mobbed or coerced by males. They simply perform and leave the decision to her. The fact that so few males mate means that females are remarkably unanimous in their choices. What is it that all prefer and can identify in a desirable male? Exactly what do females choose, and what, if anything, do they get out of their choices?

Conventional wisdom said that leks offered females an opportunity to compare males and thereby spot the superior one—a sort of contest that harkens back to the jousts of medieval knights, in which each vied to prove his superiority before the spectators. Although the males in a lek have no particular resources to offer visiting females, they were thought to at least be offering them "good" genes, as reflected in the vitality of a particular male's display or his superior location in the arena. Some biologists, however, have begun to doubt the fundamental assumption that there must always be an underpinning of "rationality" in the designs of nature. They suggest that the females may simply be "dazzled" into choosing the male who has invented the most elaborate eye- or ear-catching performance. In other words, she is influenced by slick advertising to choose a male of good form rather than proven substance.

The possibility that the elaborate displays may have evolved without any real benefit to either sex is at the center of one of the most perplexing issues with which evolutionists are currently grappling. While lek species such as the hammer-headed bat are fascinating in themselves, being such extreme cases of female choice, they also offer the best hope for answering this basic question.

As is the case among the elephant seals, the environment of the hammer-headed bat also appears to play a major part in shaping their polygamous behavior. Because the tropical fruit trees that provide the bulk of their diet are widely scattered and bear unpredictably, they can't be monopolized by a few males who could then monopolize females as they come to feed. What's more, since female hammer-heads bear and nurse their young in a largely predator-free environment, they are able to raise their young without assistance, leaving the male free, in effect, to invest his time in something else—namely, in mating with as many other females as possible.

One of the smaller flying foxes, the Gambian epauletted bat, adds an interesting twist to the courtship repertoire of the genus. With a courtship song reminiscent of the hammer-headed bat, epauletted males gather at dusk, inflate elastic cheek pouches, and emit a singsong honking sound audible for 200 yards or more. The male then unfurls tufts of white hair from pockets on each shoulder—hence the name. These epaulets are thought to contain pheromones. Wafted off into twilight by the male's gently flapping wings, they may help to attract a female.

One of the few naturalists to witness this behavior is Merlin Tuttle, a bat scientist, photographer, and publicist. He is also the founder and director of Bat Conservation International. Centered at the Brackenridge Zoology Field Laboratory at the University of Texas, Austin, it has 1,500 members in thirty-three countries working for the bat's preservation. Tuttle is also the only person to have photographed the remarkable courtship of epauletted bats.

After several unsuccessful attempts to photograph the bats in the tropical forests, Tuttle realized that they frequented the lampposts along the main street of Kisumu, Kenya, a town on Lake Victoria. These "townies" had the advantage of being able to display all night, long after their wilderness counterparts had been forced by darkness to retire. Others gathered in the trees along the main street, gaining the benefit of the lights, but foregoing the dangers of total exposure.

According to Tuttle, the males arrive at about dusk, and spend the evening defending their sites against other males. The females arrive "fashionably late," at about 11 P.M. They hover a foot or two in front of a male, who displays his epaulets and begins to accelerate his honking. Tuttle calculated that in an eight-hour courting period, each male may make more than 26,000 calls and 100,000 wing beats. Not surprisingly, most retire exhausted by 3 A.M.

It is ironic that so maligned a creature is actually responsible for some of the most intriguing courtship behavior. Shy, gentle, and good natured, the bats were easily taught to accept food from Tuttle's hands. He even gathered several in mist nets and brought them back to his hotel for an extended photo session, in which he could capture their behavior under precisely controlled lighting. As it happens, the bats also make exceedingly good mothers. A female will often fly carrying a pup, which clutches her belly while it sleeps or nurses—a cargo that may weigh two-thirds her own weight. Tuttle has even seen a pup follow its mother as she leaves the roost to feed at night, compete with her for the nectar-rich blossoms of a baobab tree, then tuck itself against her tummy and nurse during the return flight home.

However much humans equate monogamy or polygamy with moral or hedonistic choice, mating arrangements, for the large majority of mammals, are a consequence of necessity. It would be a

155

mistake, however, to strictly apply the ecological principles of other mammals to humans, for human pair-bonding involves many social and moral variables. In one cross-cultural survey, 708 out of 849 human societies were found to be polygamous, 137 monogamous, while only four practiced polyandry—the taking of several mates by a female. But even in polygamous human societies there is a strong, often prolonged (though not necessarily exclusive) alliance between mates. What's more, only human societies offer the ceremonious legitimization of relationships and of offspring. Unlike most other mammals, human children, virtually universally, need dual parental investment in their rearing.

It has been speculated that the importance of the father in raising the human infant is closely intertwined with the momentous evolutionary change that enabled our ancestors to walk upright. Anatomically, this may have led to a narrowing of the birth canal. Birth, in turn, became a difficult and complicated matter. Natural selection began to favor delivery of smaller, lesser-developed offspring that could easily pass through the canal. Because such offspring required years of diligent parental care, human fathers found themselves with a permanent, essential role to play in the raising of offspring.

12.

In the Image
of Man

I<small>T</small> I<small>S</small> 2:30 P.M.—feeding time at the field station of the Yerkes Regional Primate Research Center, Emory University. Located in the clear, fresh country air near the small rural town of Lawrenceville, Georgia, some twenty miles from metropolitan Atlanta, the station features chimps' quarters that consist of a long rectangular pen constructed of heavy wire mesh, with an enclosed den at one end. Hemp ropes hang from the ceiling of the enclosure like liana vines, and the gradually sloping cement floor is covered with fresh straw. On this chilly November day the chimps have retired to the warmth of their den, but as the dominant male, Bozondjo, hears the rattling of carrot-, orange- and cabbage-filled aluminum buckets carried by the keeper walking down a grassy slope toward the pen, he steps outside, followed by the rest of his troop—three adult females and a female infant: Linda; the dominant Laura; her two-year-old infant, Lisa; and Lorel. As they spot the khaki-garbed figure of the keeper, they explode into a cacophony of shrieks and cries. Bozondjo is fifteen years old, with an alert, powerful body shrouded in lush black fur, a wrinkled face, and hands and feet of shiny black leather. In a slouch, with knuckles braced against the

157

floor, he stands about seventy-five centimeters tall at the shoulders. He is more graceful than a common male chimpanzee, his face more expressive, and his body more lithe and dynamic. As he spots the food, Bozondjo's slender pink penis immediately becomes erect and conspicuous.

He quickly grabs several carrots and oranges as they are pushed through the wire. Gripping a large orange in each hand, he turns around and displays for Linda, the eldest of the group. Shy, retiring, and forever living in the shadow of the domineering Bozondjo, she avoids eye contact and instead immediately turns around and "presents" her rump for a copulation, while gazing under her arm at him. After copulating, Bozondjo ambles across the enclosure to the dominant female, Laura, swaggering toward her in a low squat, but she ignores him. Moments later, she presents her rump. Bozondjo mounts her for a half a minute, while she gazes under her arm at him. He then stands back and swaggers as he peels his own orange by gripping the rind between his big teeth, tearing it away from the pulp, and spitting it on the floor. Linda, meanwhile, is free to grab the remainder of the food.

Bozondjo pops the last of the orange between his teeth, then lopes toward Lorel. Young, sturdy, and independent, she ignores him and continues munching on a carrot. Bozondjo squats near her with his legs akimbo, still sporting an erection. She is unimpressed, perhaps knowing that she has more choice in the outcome than the subordinate Linda. To make his presence even more conspicuous and to distract the female from the food, he arches backward to advertise his pink penis and flicks his wrists urgently up and down. His almost hairless scrotum accentuates the size of his genitalia, contrasting it with the black fur of his thighs—a clue to the importance of genital displays in the courtship of pygmy chimps.

By the time the last bucket of food is emptied, each of the chimps has sequestered a share of the food, with Laura the clear winner in amount. In addition to gripping an orange in each foot and each hand, she has a stack of celery balanced in the inside of her left elbow. Bozondjo, meanwhile, momentarily retires to the end of the enclosure and slowly munches on his food. The infant Lisa, not yet weaned, chews ineffectually on the tip of a celery stalk. Their stomachs full, and their ranks and alliances strengthened by the pattern of their sexual activity, the chimps retire to the warmth of the den.

After a short rest, Bozondjo lopes across the floor in a low crouch, swaggers his shoulders side to side as his long arms hang nearly to the ground and swing like hairy pendulums. He approaches Laura in a low, confident swagger, then holds out his hand and offers her some of his food. Failing to gain her attention, he nudges her with his elbow. Laura shifts her shoulders and glances at him out of the corners of her eyes, and finally turns toward him as if to ask, "What do you want now?"

With a circular motion of his right index finger, Bozondjo signals her to turn around. Laura refuses, but instead rolls back on her haunches while facing him, and spreads her legs. She, like most female pygmy chimpanzees, prefers to mate face to face— possibly more pleasurable for her than the male-preferred front to back because, when the genitals are swollen, the female chimp's clitoris points forward between her legs and is easily stimulated by mating front to front. Although functionally similar to that of women, this configuration is unique among the great apes.

Bozondjo persists, signaling Laura to turn around. Finally she does, but as Bozondjo begins to mount, she swings back around, flops down on her haunches, and spreads her legs. She is apparently trying to trick him into face to face copulation. Slightly annoyed, Bozondjo flicks his wrists and fingers in a general signal, as if conveying the message, "Shape up!" Laura again turns and presents, but as the frustrated Bozondjo tries to mount, she suddenly swings around again and spreads her muscular legs. Finally she gives in. Her lips part into a wide grimace and she presents dorsally as Bozondjo mounts, and she emits a loud *eeeeeeeeep!* Bozondjo smacks his lips as Laura reaches between her legs and affectionately holds his scrotum, pulling him against her thighs as he thrusts. As he dismounts, Laura quickly turns around, leans back on her bony knuckles as if beckoning him to "come hither." Bozondjo does not resist but mounts by pulling her up and onto him by holding her around the waist. They mate face to face, occasionally gazing into each other's eyes.

Bozondjo is a good ape, a vigilant father, and a protective mate. Much to Laura's satisfaction, he also likes to baby-sit, frequently spelling her from the arduous demands of tending their infant Lisa, born in 1981. As Laura sits in one corner grooming the neck and shoulders of the suckling infant, Bozondjo strolls over. Lisa drops the teat from her mouth and looks up. Bozondjo plops down on the ground and holds his arms outward and upward, looks at Laura, and turns his head toward the back wall as if to say, "How about going over there and leaving Lisa with me." After some hesitation, Laura stands up and hands over Lisa. In a gentle, avuncular manner, he carries the baby across the enclosure, then lies back, placing her carefully on his lower belly, facing him. He begins to bounce it up and down as if pretending to copulate. He quivers and shudders and appears to enjoy the activity, though he does not exhibit an erection.

Soon Bozondjo grows tired of the play. He puts Lisa down in the straw and spreads it around her as if building a nest, but the infant jumps up and runs into the den. No sooner has Bozondjo relaxed in the soft bed of straw than a small furry head with large dark eyes peers out of the doorway, followed by the rest of Lisa. She rushes out and leaps on Bozondjo's back before making a quick escape to the safety of her grandmother, Linda, who sits inside. Bozondjo leaps up playfully and chases after the squealing youngster. He grabs Lisa and playfully drags her back outside. Again she bounces from his arms and escapes, with Bozondjo in hot pursuit. Lisa, looking for a little protection from a game that is getting out of hand, runs toward her mother, Laura, sitting in the interior. By this time Laura is a little alarmed by Lisa's perturbed state and faces the approaching Bozondjo. She squats bipedally, extends her arms down, places her wrists together with her hands facing out, and gives a stern pushing movement toward Bozondjo, as if to say, "Now cut it out!" Bozondjo stops dead in his tracks, apparently recognizing the meaning of Laura's gesture. He sits quietly until things have cooled down. Once Lisa has gathered her composure, she initiates a new bout of play with her father who, more gently this time, commences the game of catch-as-catch-can again.

Momentarily relieved of maternal obligations, Laura sees Lorel gazing at her from across the pen. Lying backward, she extends her legs forward as Lorel maneuvers toward her. They gaze intently at each other. Then Lorel straddles Laura's lap, wraps her legs around her waist, and drops her pelvis between Laura's spread thighs until their genitals touch. In rapid, rhythmic movements, they begin rubbing them together. Laura grimaces and lets out an *EEEE! EEE! EEE!*

Drawn to the activity, Bozondjo comes over to watch. Clearly excited, his slender penis emerges. And as Laura and Lorel separate a few moments later, he reaches over and begins to stroke Lorel's genitals with his fingers, as the infant Lisa clutches to his belly. But hungry and disturbed by the commotion, she begins to shriek. The mother, Laura, immediately stands up and reaches for the infant and bundles her against her breast. Lisa sucks contentedly. Bozondjo, meanwhile, meanders to a far corner of the enclosure and, standing alone, nibbles a piece of celery.

*   *   *

Pygmy chimpanzees, gentle primates native to the primal rain forests of Zaire, in central West Africa, are among the rarest and most intelligent of the great apes. Linda, Lorel, Bozondjo, and Lisa are among the few in captivity. It is often said that the more intelligent a species is, the less ritualized its courtship becomes. So it appears with the pygmy chimps. As this portrait shows, their courtship is not particularly dazzling. Indeed, like human beings, the chimps communicate by gestures whose subtlety often belies their complexity: a glance here, a hairy nudge there, a soft touch with a wrinkled brown hand, or a supple arm placed around the shoulders of a mate. Upon meeting one another after an absence, they sometimes extend their hands in greeting.

This composite portrait of the sexual behavior of the pygmy chimpanzee has been loosely drawn from the research of Jeremy Dahl, a physical anthropologist at the Yerkes Regional Primate Research Center near Atlanta. This has included some 600 hours of highly detailed observations of both common and pygmy chimps between 1981 and 1986. Some key examples of socio-sexual interaction have been documented using video tape, still photography, and audio tape recordings. While the chimps mentioned are real residents of the Lawrenceville facility, the behaviors did not necessarily occur in the sequence described. This sketch has been supplemented by observations made by the few zoologists who have undertaken the difficult task of observing the animals in the wild, most notably Shehisa Kuroda of Kyoto University, in Japan, and Randall Sussman and students at the State University of New York, Stonybrook.

The point is that sexuality permeates nearly every activity of the species. Unlike lower animals, courtship and mating among chimps are not reserved for a particular place or season. As among humans, sex is at the core of behavior in general. In the case of the chimps, it is central to the bonding of females with females, an essential part of adult-infant relationships, and at the heart of food sharing. Driven by hunger, fear, and, perhaps above all else, sex, the chimps might approximate Sigmund Freud's idea of the id—the lusty, craving, driving force behind the veneer of human civilization.

Probably at no time are the subtle group and individual dynamics of the pygmy group more visible than at feeding occasions. Among wild pygmy chimps, some ninety percent of copulations occur at feeding time, with the majority occurring when the troop arrives at the first feeding session at the beginning of the day. In captivity, the story is similar: Bozondjo no sooner gathers an armload of food than he begins to use it as an enticement in his sexual displays toward females. This may include an indication of a readiness to share food and probably has longer-term significance: food sharing may be a way of appraising a male's competence as a provident father and mate. Like Bozondjo, a male will often hold out food to a female as if to say, "See, I'm willing to share this with you." He may even share food with an infant after copulating with females affiliated with its mother.

Choice of a provident father or of some closely affiliated females may be essential because one of the major causes of infant death among wild common chimps is malnutrition. Females who mate with males of proven generosity are less likely to have offspring who suffer from lack of food. Furthermore, if the male is willing to go out and collect food, it not only spares the female the dangers of hunting, but also enables her to spend more time with her infant—an essential requirement for raising a healthy one. This food sharing among the chimps may even shed some light on the evolution of the sexual division of labor in those human societies (the !Kung bushmen are among the exceptions) where males hunt or gather, while females remain behind to protect the young, then share in the food when the males return.

Among common chimps, food sharing is sometimes initiated

by begging, with a lower ranking chimpanzee extending his hand and requesting a handout. The urge to attack the would-be usurper is often undermined by the submissive gesture—a timid, outstretched hand and a nonthreatening gaze—a gesture similar to the outstretched arm that chimps sometimes display upon greeting one another. In fact, it has been speculated that the human custom of extending a hand to strangers has evolved from a similar gesture of appeasement, evolutionarily rooted in food begging.

The sexual behavior of chimps is not only at the center of food sharing, but may also play a role in building trust among individuals, especially among females. In both captive and wild animals, female "affiliation" is often expressed by rubbing genitals. This may relieve the tension or stress, and may occur after a female has shared some of her food with a beggar. Sometimes the approach of a human stranger will send captive females into a bout of genital-genital contact. It may also be an important outlet for reaffirming bonds among lactating females, occurring as it often does among captive chimps when a male volunteers to baby-sit, thus freeing mothers from their maternal duties and permitting them to seek the intimate company of other females. In a larger sense, this sharing of a mutually pleasurable activity may strengthen overall cohesion of a group by promoting friendly and trusting relationships between females. In a maternally based society—that is, with females forming the core of a troop—strong female-female alliances are crucial.

Sex not only cements bonds between females, it also plays an important role in the development of bonds between adults and infants. While baby-sitting for a female between nursings, an adult male will engage in sexual play with the infant. The fact that males are around and willing to baby-sit, and that the females willingly seize the opportunity to stimulate each other, provides the basis for an on-going process of cementing social bonds—infants with adults and females with other females.

By most human standards, the activities engaged in by Bozondjo and his infant (and the genital-rubbing between an infant and its mother or grandmother) would be called child abuse. In fact, this sort of behavior may play an important role in the development of the infant chimp's own sexuality. Some researchers suggest that this facet of chimp behavior might offer some evolutionary insight into the problem of child abuse in our own society. Might such child abuse be the expression of an evolutionary tendency which, though serving a useful purpose among our closest relatives, has resurfaced with a destructive persistence among *Homo sapiens*? "Current knowledge of the behavior of pygmy chimps may provide a new perspective on infant-adult sexual relations as well as on the nature of female homosexual behavior," says Dahl. "Still, in the end, a new perspective raises as many questions as it answers."

In the primates, one can see to a degree found in no other nonhuman animal, the emergence of a new meaning of sexuality. A male chimp's courtship patterns are not triggered only at the female's peak fertility period, when copulation has a good chance of resulting in pregnancy. Females can be attractive to males even during the long periods when they are not capable of conceiving. In many nonhuman primates, one can see the dawning of sex for nonprocreative purposes: social bonding, the release of tension, bribery. This nonreproductive sexual activity appears to reach an unusually high level in pygmy chimps. This is a startling revelation since, until recently, frequent nonreproductive sex was thought to be a uniquely human characteristic.

Like the behavior of humans themselves, the courtship of pygmy chimps is largely based on an amalgam of learning and experience. The complex texture of their behavior becomes most glaringly apparent when it is compared to those of animals at the opposite end of the evolutionary scale—the horseshoe crab and grunion, for example—for whom courtship is little more than a linear progression of genetically programmed events. Looking at the relatively simple grunion, one is nonetheless amazed that it has mastered such complexities of behavior as catching the precise tide to shore. Looking at the pygmy chimp, one is tempted to believe it is as aware of an observer's existence as the observer is of the chimp's. Indeed, like humans, pygmy chimps seem to take great pleasure if not actual solace in their sexual relations. Between the evolutionary extremes, one finds an almost endless series of courtship arrangements. Yet all are directed in large degree toward the same end: creating new life.

Were the history of life reduced to the scale of an hour, the primate would have arrived in the last ten seconds. The mere fact that they practice reproductive sex at all marks them as relatively recent arrivals, for, as noted earlier, sexual reproduction was a very late invention in evolutionary terms.

Sexual reproduction has obviously served the interests of many animals well, but its "modernness" should not be confused with efficiency. As even a cursory view of courtship shows, sex is expensive, in both genetic and energetic terms. While one is apt to envision courtship as among the simpler and more basic aspects of nature, it is actually one of the most diverse and biologically complex.

The pygmy chimpanzee, which shares about ninety-nine percent of the same genetic material as humans, may be the most humanlike of the great apes in many aspects of behavior. Consequently, in them one begins to see the most compelling parallels between animal behavior and the roots of our own.

But however close the parallels, the chimp offers, at most, insight only into where we came from—our evolutionary roots—not who we are, per se. Animals are not a mirror image, but merely a shadow, of ourselves. As far as that goes, why stop with the pygmy chimps? We can find semblances of human behavior practically all the way down to the horseshoe crab, if only one looks long enough.

Hundreds of millions of years ago, in the warm waters off some dark and silent continent, when sex and courtship began to evolve, the world was marked with a potential for the unfathomable complexity that surrounds us today. In tracing this bounty of nature from its ancient beginnings, we are apt to forget that life today may mark the beginning of countless evolutionary epochs to come. Life is forever evolving, patterning itself for survival.

Source Notes

CHAPTER ONE: ANCIENT BEGINNINGS

1: Carboniferous Period: Kummel, Bernhard. 1970. Pp. 246–252.

4: Five kingdoms of living things: Life has traditionally been partitioned into two or three kingdoms. Recent advances in the understanding of evolutionary biology, however, have led some scientists to adopt a five-kingdom schema.

9–10: Hamilton's theory: Wertheim, Margaret, 1986.

11: Corn Mother and Indian maiden legends: Berrill, N.J. 1953. P. 5

11: Pawnee and Zuni Legends: Haeberlin, H.K. 1916.

11–12: Afro-American spider legend: Abrahams, Roger D. (ed.) 1985. P. 45.

13: "A forty-four-pound ling . . . carried 28,361,000 eggs": Grzimek, Bernhard. 1973. P. 64.

14–15: Black hamlet: Fisher, Eric A. 1980.

16–17: Two-lined salamander: Halliday, Tim. 1980. P. 109.

17: Tailed frog: *ibid.* P. 119.

20–21: Similarities between human blood and sea water: Carson, Rachel L. 1951. Pp. 13–14.

21: Six percent of human newborns: *Biology Today.* 1972. P. 867.

21: Boy with a nine-inch tail: Schultz, A.H. 1926.

CHAPTER TWO: THREE LUNAR DANCES

35: Peak time for onset of labor: Palmer, John D. 1976. P. 153.

35: Human sensitivity to house dust: *ibid.* P. 156.

36: Story of J.X.: Miles, L.E.M., Raynal, D.M., and Wilson, M.A. 1977.

37–38: Columbus and Bermuda fireworm: Cloudsley-Thompson, J.L. 1961. Pp. 81–82.

39: Galloway visit to Bermuda: Galloway, T.W. 1908.

45–46: Crane experiment with wiggling stick: Crane, Jules M., Jr., 1969.

48: Full moon and conceptions: Catton, Chris, and James Gray. 1985. Pp. 216–217.

CHAPTER THREE: THE RISE OF CONFLICT

50: Anthansius Kircher: Browne, Janet. 1983. P. 4.

50: "In 1975 alone, 786 . . . species of flies were described": Ehrlich, Paul R., and Anne. 1981. P. 17.

52: Darwin quote: Darwin, Charles. 1901.

52: Trivers quote: Trivers, Robert. 1972.

53: "'Sex is an antisocial force in evolution'": Wilson, Edward O. 1975. P. 314.

53: "The human ovum is 85,000 times larger than the sperm": Wilson, Edward O. 1978. P. 97.

55: European smooth newt: Halliday, Tim. 1980. Pp. 109–110.

57: Praying mantis: Bermant, Gordon, and Julian M. Davidson. 1974. Pp. 103–106.

59: *A Herd of Red Deer:* Darling, Frank F. 1969. Pp. 160–161.

60: Violent or deadly combat in hippopotami, musk oxen, grizzly bears, and langurs: Southwick, Charles H. 1970. P. 4.

60: Fossey studies of mountain gorillas: Fossey, Dian. 1983.

61–62: Story of Mug and Itch: Hrdy, Sarah Blaffer. 1977. Pp. 242–290.

CHAPTER FOUR: SPRING

71: Cross study of "an aggressively frigid female": Austin, C.R., and R.V.B. Short, (eds.) 1972. P. 40.

73: Experiments with testosterone: de Kruif, Paul. 1945.

74: "After resounding victories testosterone levels rose . . .": Svare, Bruce B. (ed.). 1983. P. 570.

75: Animals without pineal glands: Mess, B., Ruzsas, C.S., Tima, L., and Pevet, P. (eds.). 1985. P. 165.

76: "Assault and rape peaked": Herbert, W. 1983.

CHAPTER FIVE: TOWARD A PEACEABLE KINGDOM

78: "Chaffinches . . . flying at forty-four kilometers an hour": Welty, Joel C. 1975. P. 477.

79: Swallows demand 15 centimeters . . .": Wilson, Edward O. 1975. P. 257.

80–84: Chaffinch courtship: Hinde, R.A. 1953.

84–85: "Among birds the most aggressive . . .": Lorenz, Konrad. 1964. P. 48.

85: Cichlids: *ibid*. Pp. 45–47.

87–88: Spotted Wolf and Piebald Eagle: Lorenz, K. 1966. Pp. 73–64.

88–91: Black-headed gull: Moynihan, M. 1955.

91: Courtship as mix of aggression and submission: Southwick, Charles (ed.). 1970. Pp. 28–29.

91: Edward Armstrong's phrase: Wilson, Edward O. 1975. P. 225.

91: Similarity between courtship "appeasement gestures" and limits on male conflict: Tinbergen, Niko. 1974. P. 20.

92: Signals that shorten fights: Halliday, Tim. 1980. Pp. 93–96.

96: Arabian oryx: Smith, Maynard, and G.R. Price. 1973. P. 15.

96–97: Curt Sachs quote: Sachs, Curt. 1952.
97: Pondo women: Hunter, Monica. 1961. P. 408.

CHAPTER SIX: SONG

98–99: Wandering spider: Rovner, Jerome S., and Friedrich G.
Barth. 1981.
98: *Aedes Vexans:* Frings, Hubert and Mable. 1964. Pp. 1–2.
101: Electric fish: Hopkins, Carl D. 1977. Pp. 263–289.
101: "The wailing of . . . spadefoot toads": Wilson, Edward O.
1975. P. 443.
104: Honey guide: Lanyon, W.E., and W.N. Tavolga. 1960. Pp.
371–372.
106: It was reported by Patrick O'Donovan in the *Observer* on June
12, 1960, that immediately after the Chilean earthquake, birds
burst out singing "loudly and discordantly": Armstrong, Edward A. 1963. P. 204.
113: Tyack quote: Bright, Michael. 1984. P. 28.
125–126: Reported sightings of whales mating: Slijper, Everhard.
1979.
127: Darwin's quote on language and courtship: Darwin, Charles.
1901. P. 133.

CHAPTER SEVEN: SILENT COMMUNICATION

129–130: History of gypsy moth in America: Gerardi, Michael H.,
and James K. Grimm. 1979.
131: "The male's elaborate antennae": Schneider, D. 1974. Pp.
28–35.
132: Courtship and mating of gypsy moth: Gerardi, Michael H.,
and James K. Grimm. 1979.
134: "urchins gathered . . . to marvel at the convention of moths":
Schneider, D. 1974. Pp. 28–35.

134: "Pheromonal air assault": Beroza, M. 1976.

138–139: "'Boar Mate' used to render sows more amenable to artificial insemination": Vandenbergh, John G. 1983. P. 262.

139: Description of pig courtship: Hafez, E.S.E. (ed.). 1975. Pp. 302–305.

140: "The biological role of this boar sex pheromone . . .": Claus, R., Hooper, H.O., and Karg, H. 1981.

143: "an aphrodisiac of coconut oil mixed with turmeric": Beach, Frank A. (ed.). 1974. P. 184.

150: "McClintock Effect": McClintock, Martha K. 1971.

CHAPTER EIGHT: UNCONVENTIONAL COURTSHIP

154–165: "Alternative strategies" for gaining "access to females": Alcock, John. 1979. Pp. 220–231.

157–158: Cottonmouths: Martin, David L. 1984.

157: Snake "courtship ritual": Carpenter, C.C. 1977.

160–161: *Anolis garmani* lizard: Trivers, Robert L. 1974.

162: Red-sided garter snake: Mason, Robert T., and David Crews. 1985.

162: "The male stickleback fish will sometimes mimic the behavior of a female": Assem, J. van den. 1967.

163–164: Carnivorous *Photuris* firefly: Lloyd, James E. 1965.

167: "'The cutting edge of evolution'": Wilson, Edward O. 1975. P. 290.

167: "Evolution of the fourteen species of finches in the Galapagos Islands": Christian, John J. 1970.

CHAPTER NINE: THE MONOGAMOUS WAY OF BIRDS

182: Kittiwake gulls: Coulson, J.C. 1966.

182: Perching songbirds: Welty, Joel C. 1975. P. 248.

183: Blue herons and great egrets of Hog Island, Texas: Diamond, Jared. 1985.

184–185: Pied flycatcher: Alatalo, Rauno V., Carlson, A., Lundberg, A., and Ulfstrand, S. 1984.

187–188: Red-necked phalarope: Tinbergen, Niko. 1935.

CHAPTER TEN: THE PRICE OF PARADISE

197–198: Courtship of magnificent bird of paradise: Rand, A.L. 1940.

199: Great-tailed grackle: Orians, G.H. 1969. P. 601.

CHAPTER ELEVEN: WHY MAMMALS ARE UNFAITHFUL

204: "Guinea pig is . . . sufficiently well developed at birth to survive without milk": Cowie, A.T., Forsyth, J.A., and Hart, J.C. 1980. P. 1.

204: "Pilot whale calf may suckle for up to seventeen years": Kevles, Bettyann. 1986. P. 136.

206: The influence of the human breast: Wickler, Wolfgang. 1972. Pp. 246–254.

209: Klipspringer: Dunbar, Robin. 1985.

219–221: Hammer-headed bat: Bradbury, Jack W. 1977.

228: Survey of 849 human societies: Daly, Martin, and Margo Wilson. 1983. P. 265.

CHAPTER TWELVE: IN THE IMAGE OF MAN

Material for this chapter came almost exclusively from the author's personal conversations with the authorities cited in the text.

Bibliography

Abrahams, Roger D. (ed.) 1985. *Afro-American Folktales: Stories from Black Tradition in the New World*. New York: Pantheon Books.

Atalato, Rauno V., Carlson, A., Lundberg, A., and Ulfstrand, S. 1981. "The Conflict Between Male Polygamy and Female Monogamy: The Case of the Pied Flycatcher *Ficedula hypoleuca*," *American Naturalist 117*: 738–753.

Alatalo, Rauno V. *et al.* 1984. "Male Deception or Female Choice in the Pied Flycatcher *Ficedula hypoleuca:* A reply," *American Naturalist 123*: 282–285.

Alcock, John. 1979. *Animal Behavior: An Evolutionary Approach*, 2nd. ed. Sunderland, Mass.: Sinauer.

Armstrong, Edward A. 1963. *A Study of Bird Song*. New York: Oxford University Press.

Assem, J. Van den. 1967. "Territoriality in the three-spined Stickleback, *Gasteroceus aculeatus*," *Behaviour Supplements* 16:1–164.

Austin, C.R., and R.V. Short (eds.). 1972. *Reproduction in Mammals: Hormones in Reproduction*. London: Cambridge University Press.

———— 1979. *Reproduction in Mammals: Mechanisms of Hormone Action*. London: Cambridge University Press.

Barash, David P. 1977. "Sociobiology of Rape in Mallards *(Anas platyrhynchos):* Responses of the Mate Male," *Science 197*: 788–789.

———— 1982. *Sociobiology and Behavior*, New York: Elsevier.

Bibliography

Barlow, Robert B., Jr., L.C. Ireland, and L. Kass. 1982. "Vision Has a Role in *Limulus* Mating Behaviour," *Nature 296* (March 4):65.

Barlow, Robert B., Jr. 1983. "Circadian Rhythms in the Limulus Visual System," *The Journal of Neuroscience 3* (April): 856–870.

Barlow, Robert B., Jr., Maureen K. Powers, Heidi Howard, and Leonard Kass. *Migration of Limulus:* Relation to Lunar Phase, Tide Height and Sunlight." 1986. *Biological Bulletin* 171:310–329.

Bartholomew, George A. 1970. "A Model for the Evolution of Pinniped Polygyny," *Evolution 24:* 546–559.

Beach, Frank A. (ed.). 1974. *Sex and Behavior.* Huntington, New York: Krieger.

Benton, David. 1982. "The Influence of Androstenol—a Putative Human Pheromone—On Mood Throughout the Menstrual Cycle," *Biological Psychology 15:* 249–256.

Bermant, Gordon, and Julian M. Davidson. 1974. *Biological Basis of Sexual Behavior.* New York: Harper and Row.

Beroza, M., and E.F. Knipling. 1972. "Gypsy Moth Control with the Sex Attractant Pheromone," *Science 177 (July):* 19–27.

Beroza, M. 1976. "Control of the Gypsy Moth and Other Insects with Behavior Controlling Chemicals" in *Pest Management with Insect Attractants and Other Behavior Controlling Chemicals,* M. Beroza (ed.). Washington, D.C.: American Chemical Society. Pp. 99–118.

Berrill, N.J. 1953. *Sex and the Nature of Things.* New York: Dodd, Mead.

Bibson, R.W., and J.A. Pickett. 1983. "Wild Potato Repels Aphids by Release of Aphid Alarm Pheromone," *Nature 302:* 608–609.

Biderman, John O., and David C. Twichell. 1983. "Food for Flight," *Audubon 85* (May):112–119.

Birch, Martin C. (ed.). 1974. *Pheromones.* New York: Elsevier.

Botton, Mark L., and Harold H. Haskin. 1984. "Distribution and Feeding of the Horseshoe Crab, *Limulus polyphemus,* on the Continental Shelf off New Jersey," *Fishery Bulletin 82* (2): 383–389.

Bradbury, Jack W., 1977. "Lek Mating Behavior in the Hammer-Headed bat," *Zeitschrift fur Tierpsychologie 45:* 225–255.

Bright, Michael. 1984. *Animal Language.* Ithaca, New York: Cornell University Press.

Brown, Frank A., Jr., J. Woodland Hastings, John D. Palmer. 1973. *The Biological Clock: Two Views.* New York: Academic Press.

Browne, Janet. 1983. *The Secular Ark: Studies in the History of Biogeography.* New Haven: Yale University Press.

Bunning, Erwin. 1973. *The Physiological Clock; Circadian Rhythms and Biological Chrometry.* New York: Springer-Verlag.

Burns, Jeffrey T., Kimberly M. Cheng, and Frank McKinney. 1980. "Forced Copulation in Captive Mallards," *Auk 97:* 875–879.

Busnel, R.G. (ed.). 1963. *Acoustic Behavior of Animals,* London: Elsevier.

Carpenter, C.C. 1977. "Communication and Displays of Snakes," *American Zoologist 17:* 217–223.

Carson, Rachel L. 1951. *The Sea Around Us.* New York: Oxford University Press.

Catton, Chris, and James Gray. 1985. *Sex and Nature.* New York: Facts on File.

Cheny, D.L., and R.M. Seyfarth, 1977. "Behavior of Adult and Immature Baboons during Inter-group Encounters," *Nature 269:* 404–406.

Christian, John J. 1970. "Social Subordination, Population Density, and Mammalian Evolution," *Science 168:* 84–90.

Claus, R., H.O. Hoppen, H. Karg. 1981. "The Secret of Truffles. A Steroidal Pheromone?" *Experientia 37:* 1178–1179.

Cloudsley-Thompson, J.L. 1961. *Animal Behavior.* New York: Macmillan.

Cohen, James A., and H. Jane Brockmann. 1983. "Breeding Activity and Mate Selection in the Horseshoe Crab, *Limulus Polyphemus," Bulletin of Marine Science 33* (2): 274–281.

Colquhoun, W.P. (ed.). 1971. *Biological Rhythms and Human Performance.* New York: Academic Press.

Comfort, Alex. 1971. "Likelihood of Human Pheromones," *Nature 230* (April 16): 432–479.

Constanz, G.D. 1975. "Behavioral Ecology of Mating in the Male Gila Topminnow, *Poeciliopsis occidentalis* (Cyprinodontiformes: Poeciliidae)," *Ecology 56:* 966–973.

Cooke, F., *et al.* 1981. "Mate Change and Reproductive Success in the Lesser Snow Goose," *Condor 83:* 322–327.

Coulson, J.C. 1966. "The influence of the pair-bond and age on the breeding biology of the kittiwake gull *Rissa tridactyla," Journal of Animal Ecology 35* (2): 269–279.

Cowie, A.T., Isabel A. Forsyth, I.C. Hart. 1980. *Hormonal Control of Lactation.* New York: Springer-Verlag.

Cowley, J.J., A.L. Johnson and B.W.L. Brooksbank. 1977. "The Effect of Two Odorous Compounds on Performance in an Assessment-of-People Test," *Psychoneuroendocrinology* 2: 159–172.

Crane, Jules M., Jr. 1969. "The Response of Male Grunion to a Wiggling Stick," *Bulletin of Southern California Academy of Science 68* (3): 191–193.

Crews, David, and Michael C. Moore. 1986. "Evolution of Mechanisms Controlling Mating Behavior," *Science 231* (Jan. 10): 121–125.

Cutler, Winnifred Berg, George Preti, Abba Krieger, George R.

Huggins, Celso Ramon Garcia, Henry J. Lawley. "Human Axillary Secretions Influence Women's Menstrual Cycles: The Role of Donor Extract from Men," *Hormones and Behavior*. 1986. 20:463–473.

Dahl, Jeremy F. 1985. "The external genitalia of female pygmy chimpanzees," *Anatomical Record*, 211: 24–28.

——— "Sexual initiation in a captive group of Pygmy Chimpanzees *(Pan paniscus)*," *Primate Report*. In press.

Daly, Martin, and Margo Wilson. 1983. *Sex, Evolution and Behavior*, 2nd ed. Boston: Willard Grant Press.

Darling, F. Fraser. 1969. *A Herd of Red Deer: A Study in Animal Behavior*. London: Oxford University Press.

Darwin, Charles. 1901. *The Descent of Man and Selection in Relation to Sex*. John Murray: London.

Davenport, William. 1974. "Sexual Patterns and Their Regulation in a Society of the Southwest Pacific," in *Sex and Behavior*, Frank A. Beach (ed.). Huntington, New York: Krieger.

Davies, E.M., and P.D. Boersma. 1984. "Why Lionesses Copulate with More than One Male," *American Naturalist 123*: 594–611.

Dawkins, Richard. 1976. *The Selfish Gene*. New York: Oxford University Press.

de Kruif, Paul. 1945. *The Male Hormone*. New York: Harcourt, Brace.

Diamond, Jared. 1984. "Theory and Practice of Extramarital Sex," *Nature 312*: 196.

——— 1985. "Biology," *Discover*. 6 (April):71–82.

Doughty, Robin W. 1975. *Feather Fashions and Bird Preservation: A Study in Nature Protection*. Berkeley: University of California Press.

Dunbar, Robin. 1984. "The Ecology of Monogamy," *New Scientist 103* (Aug. 30):12–15.

———— 1985. "Monogamy on the Rocks," *Natural History Magazine* 94 (Nov.):41–46.

Ehrlich, Paul R. and Anne Erlich. 1981. *Extinction: The Causes and Consequences of the Disappearance of Species.* New York: Random House.

Emlen, Stephen T., and Lewis W. Oring. 1977. "Ecology, Sexual Selection, and the Evolution of Mating Systems," *Science 197:* 215–223.

Estep, Daniel Q., and Katherine E.M. Bruce. 1981. "The Concept of Rape in Non-humans; A Critique," *Animal Behavior 29:* 1272–1273.

Evered, D., and S. Clark (eds.). 1986. *Photoperiodism, Melatonin and the Pineal.* Belmont, Calif.: Pitman.

Filsinger, E.E., J.J. Braun, W.C. Monte, and D.E. Linder. 1984. "Human *(Homo sapiens)* Responses to the Pig *(Sus scrofa)* Sex Pheromone 5 Alpha-androst-16-en-3-one," *Journal of Comparative Psychology 98:* 219–222.

Fischer, Eric A. 1980. "The Relationship between Mating System and Simultaneous Hermaphroditism in the Coral Reef Fish *Hypoplectrus Nigricans (Serranidae),*" *Animal Behavior 28:* 620–633.

Fitch, Mary A., and Gary W. Shugart. 1984. "Requirements for a Mixed Reproductive Strategy in Avian Species," *American Naturalist 124:* 116–126.

Fossey, Dian. 1983. *Gorillas in the Mist.* Boston: Houghton Mifflin.

Frings, Hubert and Mabel Frings. 1964. *Animal Communication.* Waltham, Mass.: Blaisdell Publishers.

Galloway, T.W. 1908. "A Case of Phosphorescence as a Mating Adaptation," Reprinted from *School Science and Mathematics* (May). Read at first meeting of the Illinois Academy of Science at Decatur, Ill., Feb. 22, 1908.

Bibliography

Galloway, T.W., and P.S. Welch. 1911. "Studies on a phosphorescent Bermudan annelid *Odontosyllis enopla*," *Transactions of the American Micoscopical Society 30*: 13–39.

Garstka, W.R., and D. Crews. 1981. "Female Sex Pheromone in the Skin and Circulation of a Garter Snake," *Science 214 (Nov. 6)*: 681–683.

Gerardi, Michael H., and James K. Grimm. 1979. *The History, Biology, Damage, and Control of the Gypsy Moth, Porthetria dispar*. Rutherford, New Jersey: Fairleigh Dickinson University Press.

Ghiselin, Michael T. 1978. *The Economy of Nature and the Evolution of Sex*. Berkeley, Calif.: University of California Press.

Gilmore, Desmond, and Brian Cook (eds.). 1981. *Environmental Factors in Mammalian Reproduction*. Baltimore: University Park Press.

Goldfoot, D.A. 1981. "Olfaction, Sexual Behavior and the Pheromone Hypothesis in Rhesus Monkeys: A Critique," *American Zoologist 21* (1): 153–164.

Goodall, Jane. 1986. *The Chimpanzees of Gombe: Patterns of Behavior*. Cambridge, Mass.: Belknap Press/Harvard University.

Gowaty, P.A. 1983. "Male Parental Care and Apparent Monogamy Among Eastern Bluebirds *(Sialia sialis)*," *American Naturalist 121(2)*: 149–157.

Gower, D.B. 1972. "16 Unsaturated C19 Steroids. A Review of Their Chemistry, Biochemistry and Possible Physiological Role," *Journal of Steroid Biochemistry 3*: 45–103.

Graham, C.A., and W.C. McGrew. 1980. *Psychoneurendocrinology 5*: 245–252.

Grzimek, Bernhard. 1973. *Grzimek's Animal Life Encyclopedia*, Vol. 4, Fishes I. New York: Van Nostrand Reinhold.

Haeberlin, H.K. 1916. "The Idea of Fertilization in the Culture of the Pueblo Indians." American Anthropological Association. New Era Printing Co. Vol 3; no. 1. Jan.–May.

Hafez, E.S.E. (ed.). 1975. *The Behaviour of Domestic Animals.* London: Bailliere Tindall.

Halliday, Tim. 1980. *Sexual Strategy.* Chicago: University of Chicago Press.

Hamilton, W.D. 1970. "Selfish and Spiteful Behaviour in an Evolutionary Model," *Nature 228:* 1218–1220.

Hartrup, Willard W., and Jan de Wit (eds.). 1978. *Origins of Aggression.* The Hague: Mouton.

Herbert, W. 1983. "Rape Season: Legacy of Our Past?" *Science News 124* (July 23): 53.

Hinde, R.A. 1953. "The Conflict Between Drives in the Courtship and Copulation of the Chaffinch," *Behaviour 5:* 1–31.

Hopkins, Carl D. 1977. "Electric Communication," in *How Animals Communicate,* Thomas A. Sebeok (ed.). Bloomington: Indiana University Press.

Hopson, Janet L. 1979. *Scent Signals: The Silent Language of Sex.* New York: William Morrow.

Hrdy, Sarah Blaffer. 1977. *The Langurs of Abu: Female and Male Strategies of Reproduction.* Cambridge, Mass.: Harvard University Press.

——— 1981. *The Woman that Never Evolved.* Cambridge, Mass.: Harvard University Press.

Hunter, Monica. 1961. *Reaction to Conquest: Effects of Contact with Europeans on the Pondo of South Africa.* London: Oxford University Press.

Huntsman, A.G. 1948. "Odontosyllis at Bermuda and Lunar Periodicity." *Journal of the Fisheries Research Board of Canada 7:* (6).

Idyll, Clarence P. 1969. "Grunion: The Fish That Spawns on Land," *National Geographic 935 (May):* 714–723.

——— 1971. *Abyss: The Deep Sea and the Creatures That Live in It.* New York: Thomas Y. Crowell.

Jacobson, Martin, and Moreton Beroza. 1964. "Insect Attractants." *Scientific American:* 211:20–27.

Johnson, Roger N. 1972. *Aggression in Man and Animals*. Philadelphia: W.B. Saunders Co.

Kaplan, E., and R.B. Barlow, Jr. 1980. "Circadian Clock in *Limulus* Brain Increases Response and Decreases Noise of Retinal Photoreceptors," *Nature 286* (July 24): 393–395.

Kevles, Bettyann. 1986. *Females of the Species*. Cambridge, Mass.: Harvard University Press.

Kirk-Smith, M., D.A. Booth, D. Carroll, and P. Davies. 1978. "Human Social Attitudes Affected by Androstenol," *Research Communications in Psychology, Psychiatry and Behavior 3:* 379–384.

Kleinman, Devra G. 1977. "Monogamy in Mammals," *Quarterly Review of Biology 52:* 39–69.

Krames, Lester, Patricia Pliner, and Thomas Alloway (eds.). 1978. "Advances in the Study of Communication and Affect: Aggression, Dominance, and Individual Spacing." *Proceedings of the Sixth Annual Symposium on Communication and Affect*. New York: Plenum Press.

Kummel, Bernhard. 1970. *History of the Earth: An Introduction to Historical Geology*, 2nd ed. San Francisco: W.H. Freeman and Company.

Kummer, Hans. 1968. *Social Organization of Hamadryas Baboons*. University of Chicago Press.

Lack, David. 1968. *Ecological Adaptations for Breeding in Birds*. London: Methuen.

Lanyon, W.E., and W.N. Tavolga (eds.). 1960. *Animal Sounds and Communication*. Washington, D.C.: American Institute of Biological Sciences.

Le Boeuf, Burney J. 1969. "Social Status and Mating Activity in Elephant Seals," *Science 163:* 91–93.

—— 1972. "Sexual Behavior in the Northern Elephant Seal, *Mirounga angustirostris, Behavior 41:* 1–26.

—— 1974. "Male-male Competition and Reproductive Success in Elephant Seals," *American Zoologist 14:* 163–176.

—— 1978. "Sex and Evolution," in *Sex and Behavior: Status and Prospectus.* Thomas E. McGill, Donald A. Dewbury, and Benjamin J. Sachs (eds.) New York: Plenum Press.

—— 1986. "Sexual Strategies of Seals and Walruses," *New Scientist.* (Jan. 16): 36–39.

Lehrman, D.S. 1964. "The Reproductive Behavior of Ring Doves," *Scientific American 211* (Nov.): 48–54.

Lewis, Brian D., and Michael D. Gower. 1980. *Biology of Communication.* New York: John Wiley.

Lieber, Arnold L., 1978. *The Lunar Effect.* Garden City, N.Y.: Doubleday/Anchor.

Linzell, J.L. 1971. "The Role of the Mammary Glands in Reproduction," *Research in Reproduction. 3:* 2–3.

Lloyd, James E. 1965. "Aggressive Mimicry in *Photuris:* Firefly Femmes Fatales," *Science 149:* 653–654.

Lorenz, Konrad, 1964. "Ritualized Fighting," in *The Natural History of Aggression.* J.D. Carthy and F.J. Ebling (eds.). New York: Academic Press.

—— 1966. *On Aggression.* New York: Harcourt, Brace and World.

Manning, Aubrey. 1972. *An Introduction to Animal Behavior.* Reading, Mass.: Addison-Wesley.

Margulis, Lynn, and Dorion Sagan. 1986. *Origins of Sex: Three Billion Years of Genetic Recombination.* New Haven: Yale University Press.

Markert, Robert E., Betsy J. Markert, and Nancy J. Vertrees. 1961. "Lunar Periodicity in Spawning and Luminescence in *Odontosyllis enopla,*" *Ecology 42* (2): 414–415.

Marler, Peter, and William J. Hamilton III. 1967. *Mechanisms of Animal Behavior*. New York: John Wiley.

Marshall, N.B. *Explorations in the Life of Fishes*. 1971. Cambridge, Mass.: Harvard University Press.

Martin, David L. 1984. "An Instance of Sexual Defence in the Cottonmouth, *Agkistrodon piscivorus*," *Copeia 3:* 772–774.

Mason, Robert T., and David Crews. 1985. "Female Mimicry in Garter Snakes," *Nature 316:* 59–60.

Matthews, L. Harrison. 1964. "Overt Fighting in Mammals," in *The Natural History of Aggression*. J.D. Carthy and F.J. Ebling (eds.). New York: Academic Press.

McClintock, Martha K. 1971. "Menstrual Synchrony and Suppression," *Nature* (London) *229:* 244–245. (Jan. 22.)

Mess, B., C.S. Ruzsas, L. Tima, and P. Pevet (eds.). 1985. "The Pineal Gland: Current State of Pineal Research," *Developments in Endocrinology, 16.* New York: Elsevier.

Michael, R.P., R.W. Bonsall, and P. Warner. 1974. "Human Vaginal Secretions: Volatile Fatty Acid Content," *Science 186:* 1217–1219.

Michael, R.P. 1969. "The Role of Pheromones in the Communication of Primate Behavior," *Recent Advances in Primatology 1:* 101–108.

Michael, R.P., and E.B. Keverne. 1968. "Pheromones in the Communication of Sexual Status in Primates," *Nature* (London) *218:* 746–749.

Michael, R.P., and E.B. Keverne. 1970. "Primate Sex Pheromones of Vaginal Origin," *Nature* (London) *225:* 84–85.

Michael, R.P., E.B. Keverne, and R.W. Bonsall. (1971). "Pheromones: Isolation of Male Sex Attractants From a Female Primate," *Science 172:* 964–966.

Miles, L.E.M., D.M. Raynal, and M.A. Wilson. 1977. "Blind Man Living in Normal Society has Circadian Rhythms of 24.9 Hours," *Science 198:* 421–423.

Moynihan, M. 1955. *Some Aspects of Reproductive Behavior in the Black-headed gull (Larus ridibundus L.) and Related Species.* Leiden: E.J. Brill.

Myers, J.P. 1986. "Sex and Gluttony on Delaware Bay," *Natural History 95* (May):68–77.

Nadler, R.D. 1975. "Face to Face Copulation in Nonhuman Mammals," *Medical Aspects of Human Sexuality* (May): 173–174.

Neill, Wilfred T. 1971. *The Last of the Ruling Reptiles, Alligators, Crocodiles, and Their Kin.* New York: Columbia University Press.

New Scientist. 1981. "The Song of the Love-lorn Whale," 90. No. 1255 (May 28):559.

——— 1985. "How to Be a Successful Snake in the Grass." No. 1468 (August 8).

Orians, G.H. 1969. "On the Evolution of Mating Systems in Birds and Mammals," *American Naturalist 103:* 589–603.

Palmer, John D. 1976. *An Introduction to Biological Rhythms.* New York: Academic Press.

Payne, Roger (ed.). 1983. *Communication and Behavior of Whales.* American Association for the Advancement of Science/Westview Press.

Payne, Roger S., and Scott McVay. 1971. "Songs of Humpback Whales," *Science 173 (Aug. 13):* 585–597.

Payne, Roger. 1979. "Humpbacks: Their Mysterious Songs." *National Geographic* (Jan.).

Preti, George, *et al.* 1986. "Human Exillary Extracts: Analysis of Compounds from Samples which Influence Menstrual Timing," *Journal of Comparative Physiology 80:* 255–266.

——— 1987. "Human Axillary Extracts: Analysis of Compounds from Samples Which Influence Menstrual Timing," *Journal of Chemical Ecology 13* (4).

——— 1986. "Human Auxillary Secretions Influence Women's

Menstrual Cycles: The Role of Donor Extract of Females."
Hormones and Behavior. 20:474–482.

Rand, A.L. 1940. "Courtship of the Magnificent Bird of Paradise," *Natural History Magazine 45* (3):Pp. 172–175.

Reiter, Russel J. (ed.). 1984. *The Pineal Gland.* Vol. 2. New York: Raven Press.

Ripley, Dillon, and Walter A. Weber, 1950. "Strange Courtship of Birds of Paradise," *National Geographic 97*(Feb.): 247–278.

Rovner, Jerome S., and Friedrich G. Barth. 1981. "Vibratory Communication Through Living Plants by a Tropical Wandering Spider." *Science 214* (Oct. 23):464–466.

Rudloe, Jack and Anne. 1981. "The Changeless Horseshoe Crab." *National Geographic 159 (April):* 562–572.

Sachs, Curt. 1952. *World History of the Dance.* New York: Seven Arts Publishers.

Schneider, D. 1969. "Insect Olfaction: Deciphering System for Chemical Messages." *Science 163* (March): 1031–1036.

——— 1974. "The Sex-Attractant Receptor of Moths," *Scientific American 231* (July): 28–35.

Schultz, A.H. 1926. "Fetal Growth of Man and Other Primates," *Quarterly Review of Biology. 1:* 465–521.

Scott, John Paul. 1975. *Aggression.* Chicago: University of Chicago Press.

Selander, Robert K. 1972. "Sexual Selection and Dimorphism in Birds," in *Sexual Selection and the Descent of Man.* Bernard Campbell (ed.). Chicago: Aldine.

Shorey, Harry H. 1976. *Animal Communications by Pheromones.* New York: Academic Press.

——— 1977. "Pheromones," in *How Animals Communicate.* Thomas A. Sebeok (ed.). Bloomington: Indiana University Press.

Signoret, J.P. 1970. "The Mating Behavior of the Sow," *Symposium on Effect of Diseases and Stress on Reproductive Efficiency in Swine*. University of Nebraska College of Agriculture. Pp. 295–313.

Slipjer, Everhard, 1979. *Whales*. Ithaca, N.Y.: Cornell University Press.

Small, George L. 1971. *The Blue Whale*. New York: Columbia University Press.

Smith, J. Maynard, and G.R. Price. 1973. "The Logic of Animal Conflict," *Nature 246*: 15–18.

Smith, J. Maynard. 1978. *The Evolution of Sex*. Cambridge, England: Cambridge University Press.

Southwick, Charles (ed.). 1970. *Animal Aggression: Selected Readings*. New York: Van Nostrand.

Susman, Randall L. (ed.). 1984. *The Pygmy Chimpanzee: Evolutionary Biology and Behavior*. New York: Plenum Press.

Svare, Bruce B. (ed.). 1983. *Hormones and Aggressive Behavior*. New York: Plenum Press.

Tinbergen, Niko. 1935. Field Observations of East Greenland Birds. "The Behavior of the Red-necked Phalarope (Phalaropus lobatus L.) in Spring." *Ardea* 24:1–42.

———. 1958. "The Origin and Evolution of Courtship and Threat Display," in *Evolution as a Process*. Huxley, J.S., A.C. Hardy, and E.B. Ford (eds.), London: Allen & Unwin.

——— 1974. "Some Recent Studies of the Evolution of Sexual Behavior," in *Sex and Behavior*. Frank A. Beach (ed.). New York: John Wiley.

——— 1973. "On Appeasement Signals," in *The Animal In Its World: Explorations of an Ethologist*. Vol 2. Cambridge, Mass.: Harvard University Press.

Trainor, George L. 1979. "Studies on the Odontosyllis Bioluminescence System." Ph.D. dissertation. Department of Chemistry, Harvard University.

——— 1974. "Sexual selection and resource-accruing abilities in *Anolis garmani*," *Evolution 30*: 253–269.

Bibliography

Trivers, Robert L. 1972. "Parental Investment and Sexual Selection," in *Sexual Selection and the Descent of Man*. B. Campbell (ed.). Aldine: Chicago.

Tuttle, Merlin D. 1986. "Gentle Fliers of the African Night," *National Geographic*. (April).

Tyack, Peter. 1981. "Why Do Whales Sing?" *The Sciences* (Sept.):22–25.

——— 1981(a). "Interactions Between Singing Hawaiian Humpback Whales and Conspecifics Nearby," *Behavioral Ecology and Sociobiology 8*: 105–116.

——— 1983. "Differential Response of Humpback Whales, *Megaptera novaeangliae,* to Playback of Song or Social Sounds." *Behavioral Ecology and Sociogiology 13*: 49–55.

Vandenbergh, John G. (ed.). 1983. *Pheromones and Reproduction in Mammals.* New York: Academic Press.

Walker, Stephen. 1983. *Animal Thought.* Boston: Routledge & Kegan Paul.

Wallace, Alfred Russel. 1962. *The Malay Archipelago: The Land of the Orang-utan, and the Bird of Paradise.* New York: Dover.

Walser, Elizabeth Shillito. 1977. "Maternal Behavior in Mammals," in *Comparative Aspects of Lactation.* Malcolm Peaker, (ed.). New York: Academic Press.

Warner, R.R., D.R. Robertson, and E.G. Leigh, Jr. 1975. "Sex Change and Sexual Selection," *Science 190*: 633–638.

Welty, Joel C. 1975. *The Life of Birds.* Philadelphia: W.B. Saunders.

Wertheim, Margaret. 1986. "Parasites Provide a Motive for Sex," *New Scientist 109:(February 20)* 1496.

Wickler, Wolfgang. 1972. *The Sexual Code: The Social Behavior of Animals and Men.* New York: Doubleday.

Wilcox, R.S. 1972. "Communication by Surface Waves; Mating Behavior of Water Strider *(Gerridea),*" *Journal of Comparative Physiology 80*: 255–266.

Williams, George C. 1977. *Sex and Evolution*. Princeton, N.J.: Princeton University Press.

Wilson, Edward O. 1975. *Sociobiology: The New Synthesis*. Cambridge, Mass.: Belknap Press/Harvard University.

———— 1978. *On Human Nature*. Cambridge, Mass.: Harvard University Press.

Wolken, Jerome J., and Robert G. Florida. 1984. "The Eye Structure of the Bioluminescent Fireworm of Bermuda, *Odontosyllis enopla*," *Biological Bulletin 166* (February): 260–268.

Index

190